Analytical Surface Deformation Theory

for Detection of the Earth's Crust Movements

Springer
Berlin
Heidelberg
New York
Barcelona
Hong Kong
London
Milan
Paris
Singapore
Tokyo

Yüksel Altıner

Analytical Surface Deformation Theory

for Detection of the Earth's Crust Movements

With 20 Figures

Springer

Dr. Yüksel Altıner
Bundesamt für Kartographie und Geodäsie
Richard-Strauss-Allee 11
D-60598 Frankfurt/Main
Germany

E-mail: altiner@ifag.de

ISBN 3-540-65820-3 Springer-Verlag Berlin Heidelberg New York

Cataloging-in-Publication Data applied for
Die Deutsche Bibliothek - CIP-Einheitsaufnahme
Altıner, Yüksel:
Analytical surface deformation theory for detection of the Earth's crust movements/
Yüksel Altıner. - Berlin; Heidelberg; New York; Barcelona; Hong Kong; London;
Milan; Paris; Singapore; Tokyo: Springer, 1999
ISBN 3-540-65820-3

Production: ProduServ GmbH Verlagsservice, Berlin
Typesetting: Camera-ready by author
Cover design: Erich Kirchner, Springer Heidelberg
SPIN: 10717374 32/3020-5 4 3 2 1 0 - Printed on acid-free paper

To Maria, Yasemin, and Deniz Erdal

Preface

Due to plate motions, tidal effects of the Moon and the Sun, atmospheric, hydrological, ocean loading and local geological processes, and due to the rotation of the Earth, all points on the Earth's crust are subject to deformation. Global plate motion models, based on the ocean floor spreading rates, transform fault azimuths, and earthquake slip vectors, describe average plate motions for a time period of the past few million years. Therefore, the investigation of present–day tectonic activities by global plate motion models in a small area with complex movements cannot supply satisfactory results.

The contribution of space techniques [Very Long Baseline Interferometry (VLBI); Satellite Laser Ranging (SLR); Global Positioning System (GPS)] applied to the present–day deformations of the Earth's surface and plate tectonics has increased during the last 20 to 25 years. Today one is able to determine by these methods the relative motions in the cm to sub–cm–range between points far away from each other.

This high accuracy achievable using space techniques requires rethinking of the theoretical foundations of deformation analyses to be applied. These analyses constitute the subject of *chapter 2* of this book, in which an analytical theory of internal and external surface deformations is considered. Their application requires geometric surface modelling, a survey of which is given in *chapter 3*. Finally, *Chapter 4* has been dedicated to application of the model to the velocities derived within the CRODYN'94 and CRODYN'96 GPS campaigns in the Adriatic Sea area. In this case it is only a methodical application the results of which cannot be considered as sufficiently certain. More satisfactory statements can be expected only after several additional GPS campaigns in this area.

I'm grateful to *Dipl.-Ing. Branimir Gojčeta*, Director of the State Geodetic Administration in Zagreb (SGA), *Prof. Dr. Krešimir Čolić*, Geodetic Faculty of the University of Zagreb, and to *Dipl.-Ing. Aleš Seliškar*, Director of the Surveying and Mapping Authority of the Republic Slovenia in Ljubljana (SMA) for supporting the GPS campaigns in Croatia and Slovenia. Further, my grateful thanks also go to my colleagues *Dr. Božena Lipej* (SMA), *Dipl.-Ing. Zlatko Medić*, Head of the Department Land Surveying at the SGA and *Dipl.-Ing. Dušan Mišković*, Head of the Department Land Surveying at the SMA, for their useful proposals during the field work.

As the GPS data used in this book were collected with assistance of the Bundesamt für Kartographie und Geodäsie in Frankfurt/Main (BKG), I would like to express my thanks to *Prof. Dr. Hermann Seeger* and *Prof. Dr. Ewald Reinhart* for their kind support. Beyond that, I express here my gratitude to *Prof. Dr. Siegfried Heitz*, Institute for Theoretical Geodesy of the University Bonn and *Prof. Dr. Bertold Witte*, Geodetic Institute of the University Bonn for very fruitful discussions. My thanks also go to *Dr. Robert King* and *Dr. Robert Reilinger*, Massachusetts Institute of Technology, Cambridge, USA who have read and commented on certain parts of the material.

Bonn, April 1999 Yüksel Altıner

Contents

Chapter 1

Introduction

Geodesy contributes fundamentally to the geosciences by the **geometric modelling of the Earth's surface**, including the statistical as well as the time–variant representation of the internal and external geometry of the Earth's surface. The 'time–variant geometry' or the **surface deformations** are, on the one hand, of primary interest and are, beyond that, also very important as boundary values for **physical modelling of the Earth's lithosphere**.

Geodetic deformation analyses of parts of the Earth's surface have up to now been restricted nearly exclusively to the evaluation of horizontal displacements. These were generally introduced as changes of Cartesian coordinates in a mean horizontal plane and the deformation theory, based on the Cartesian coordinate geometry, was therefore conceived of as two–dimensional. Such an approach is unsatisfactory if mountainous terrain, or also large, nearly horizontal areas at the regional level are considered. If the results of a deformation analysis are to be useful as initial data based on physical or dynamic models, which generally can be assumed, they have to refer to the physical Earth's surface (topography) as seen from a mathematical point of view. This requires a representation of the Earth's surface as a generally curved surface that is embedded in three–dimensional Euclidean space. Differential–geometric modelling procedures on the basis of suitable coordinate geometries have to be applied, and for a complete treatment of surface deformations of changes of the internal geometry as well as those of the external geometry are of importance. **Geometric modelling of the Earth's surface** includes the representation of the discrete observation stations and of the observation quantities referred to them in a suitable coordinate geometry, as well as the geometric design of surface structures based on this geometry. Seen from a physical point of view, the observation stations and systems are always tied to surface points of bodies, and their mathematical representation is done on the basis of problem–oriented coordinate geometries. Rigid bodies and, to a large extent, rigid systems of rigid bodies constitute a physical and thus also a geodetic idealization, which is admissible only to a limited extent depending on the measuring accuracy as well as on the desired accuracy of the target parameters. The latter ones constitute primarily either a transformation of distance and direction measurements into a set of coordinates, or 'direct measurements of coordinates', e.g. by means of the 'Global Positioning System' (GPS). A sufficiently precise temporal invariance of the distances directly measured or computed from coordinates is the most important confirmation of a rigid point system, which is in each case limited to the period of the respective observation campaign performed. In the general case of deformable bodies moved relatively towards each other, distances are time–dependent between their surface points and these temporal distance variations are the foundation of all further **deformation theories** that start either from temporal variations of the metrics or from those of the coordinates.

From this point of view, first only deformations of body surfaces can be examined, which activity constitutes the principal portion of **regional geodetic deformation analyses**. In physics, more especially in the field of the mechanics of rigid bodies, observations of this kind of surface distortion are not an end in themselves, but on the contrary geometric boundary values as a contribution to the complete solution of problems of the dynamics of the three–dimensional mass continua of deformable bodies (Heitz 1980–1983; Heitz and Stöcker–Meier 1994). The three–dimensional deformation necessary for this has been, and is therefore, the very starting point of all deformation studies, and thus it is not surprising that in numerous textbooks it constitutes, as a rule, the sole object of consideration. **Surface deformations** are in general treated only as one aspect of three–dimensional body deformations without assigning them their own importance, which they possess without any doubt, since

- *only surface deformations can be determined directly by measuring techniques and*
- *surface deformations are often (for the time being) the only target quantities, i.e. whenever further observations for the precise determination of the material properties of the three-dimensional continua are not known, so that a complete solution of the dynamic problem is not possible.*

The geodetic community is interested in a wide spectrum of deformation phenomena. First, the global problem of the so-called **Earth tides** must be mentioned here, which concerns the deformation of the terrestrial globe owing to the tide-generating accelerations (see, e.g., Heitz 1980–1983; Melchior 1983). This problem, equally important from a geodetic as well as a geophysical point of view is still of interdisciplinary topicality, which fact holds also for the study of the stress–deformation relation of the Earth's crust on the basis of the **plate tectonics** model (see, e.g., Kersting and Welsch 1989; Miller 1992). Besides these global and large-scale regional tasks the **local analyses of surface deformations** are of utmost importance in the geodetic application, i.e. with respect to the Earth's surface (see, e.g., Pelzer 1971; Koch 1988; Saler 1995; Altıner 1996; Ghitău 1998), as well as with regard to the surface of artificial buildings, machines and human bodies (see, e.g., Caspary 1987; Reiking 1994, Bilajbegovic 1996).

The geometric-kinematic modelling of small regions of the Earth's surface are a **focal point of geodetic applications**. However, these shall also allow an insight into the causal dynamic variations of the Earth's surface. Therefore, the investigations are not restricted to coordinate or distance variations. On the contrary, special emphasis is placed on a detailed discussion in *chapter 2.3* of the **analytical theory of internal and external surface deformations**, which in accordance with the examples treated in *chapter 4* concentrates on primary coordinate determinations by means of GPS. This aspect is taken into account also by the fact that not only formulas related to Cartesian coordinates have been derived, but also the corresponding formulas on the basis of the surface normal coordinates. Here, the most important special case constitutes the ellipsoidal coordinates used by the national survey and for GPS. For two reasons the deformation theory of the three-dimensional continua precedes in *section 2.2* the surface deformation theory: on the one hand it constitutes the formal basis for the internal surface deformations, and on the other hand it is needed for dynamic or physics–based modelling. Some interesting **local connections between deformations and stresses** are illustrated in *section 2.3.4*.

In the **geodetic applications of the surface deformation theory** to the geometric modelling of deformations of the Earth's surface, as they are discussed in *section 2.3*, one has to start from **discretely distributed observation stations**

$$\{P\} \in \textit{region } F \textit{ of the Earth's surface} \qquad (1\text{–}1)$$

for which two observation campaigns by means of GPS were carried out with the mean times

$$t \textit{ and } \bar{t}. \qquad (1\text{–}2)$$

As result are then given for all P the two coordinate triplets

$$q^a := q^a(t) \,, \qquad \bar{q}^a := q^a(\bar{t}) \,, \qquad a \in \{1,2,3\} \qquad (1\text{–}3)$$

the q^a, $\bar{q}^a$ being generally ellipsoidal coordinates. From (1–3) the 'displacement coordinates'

$$z^a := \bar{q}^a - q^a \qquad (1\text{–}4)$$

can be formed. These represent the displacement field of the region F of the Earth's surface to be investigated (1–1), which is generally better the denser the field of the observation stations P is selected. It must be observed here that the station density to be selected, and thus the quality of representation of the displacement field, is not only a function of the height variations within F. Also in relatively flat areas an appropriately dense distribution of observation stations must be given to detect possible short wavelength displacement components.

Owing to (1–2, 3) $P_m - P_n$

linear elongations

$$q_{mn} := (\bar{s}_{mn} - s_{mn})/s_{mn} \qquad (1\text{–}5)$$

can be directly computed for neighbouring point pairs. This primary measure of deformation is, as a rule, not fully satisfactory. It is an 'unordered set of parameters', which does not allow a clear representation of the internal deformation behaviour in F, (1–1); and, which in particular allows no direct propositions on external surface deformations. Moreover, the linear elongation is still not a suitable starting–point for further dynamic investigations of the stress–deformation behaviour in the Earth's upper crust. All deficiencies mentioned do not apply in an analytical surface deformation theory as it is elaborated in *section 2.3.* However, this presupposes that the **discrete observation stations {P} constitute an element of a surface continuum.** In other words, the coordinates $q^a(t)$ of the surface points (1–3), and the displacement coordinates z^a (1–4) have to be modelled as analytical functions of surface coordinates u^α, $\alpha \in \{1,2\}$:

$$q^a = q^a(u^\alpha) , \qquad z^a = z^a(u^\alpha) . \tag{1–6}$$

With such a **geometric modelling in the form of interpolation or approximation functions** two properties are generally to be aimed at:

- *for the unambiguous analytical computation of the internal and external deformation measures, the interpolation functions (1–6) in the whole interpolation area shall be continuously differentiable according to the* u^α (1–7a)
- *the interpolation functions have to show an approximately linear behaviour between neighbouring observation stations.* (1–7b)

The **determination of interpolation or approximation functions** is often used in applied mathematics and informatics, which is very old and has been dealt with in great variety (see, e.g., Zienkiewicz 1972; Abramowsk and Müller 1991; Cui et al. 1992). The 'transformations of discrete function representations in analytical function spaces' have gained great importance in all technical fields including geodesy (see, e.g., Moritz and Sünkel 1977; Reiking 1994). The problem lies in the common fulfilment of both conditions (1–7a) and (1–7b), which shall be dealt with in more detail in *chapter 3.* A special case treated in *section 3.4* is the **subdivision of a surface F by finite triangle elements** or **triangulation of a surface** with linear interpolation functions (1–6) within the triangles. In this *'simplest spline–method'* only the functions themselves are continuous at the triangle sides, but not their derivatives. Thus, (1–7a) is dispensed with in favour of an exact fulfilment of the condition (1–7b). This leads to the renouncement of a completely analytical deformation theory, for which reason suitable discrete deformation measures have to be defined.

Chapter 4 is dedicated to the **fundamentals of the observation techniques** and the partial **application of the theoretical foundations** of *chapters 2 and 3.* Here, the observation campaigns and the pertinent deformations have been compiled in tables and graphs and discussed, for the Adriatic Sea area.

Chapter 2

Deformation Theories

2.1 Coordinate Systems

2.1.1 Cartesian and Curvilinear Coordinates

For the three–dimensional Euclidean geometry

Cartesian coordinates with the system of notation

$$x_{i.}\ in\ S\ , \qquad \bar{x}_{i.}\ in\ \bar{S} \tag{2.1–1}$$

are used. In three–dimensional Euclidean as well as in two–dimensional Riemannian spaces are introduced

general curvilinear coordinates:

three–dimensional:

$$q^a\ ,\quad a \in \{1,2,3\}\ , \qquad metric\ tensor\ g_{ab}\ ; \tag{2.1–2a}$$

two–dimensional surface coordinates:

$$u^\alpha\ ,\quad \alpha \in \{1,2\}\ , \qquad metric\ tensor\ f_{\alpha\beta}\ . \tag{2.1–2b}$$

The coordinate geometry on the basis of these systems is represented according to Heitz (1988).

For the coordinate geometry the

general differential equations of the coordinate transformations $q^a \Leftrightarrow \bar{q}^a$:

$$\bar{q}^c_{,ab} = \Gamma^d_{ab}\,\bar{q}^c_{,d} - \bar{\Gamma}^c_{de}\,\bar{q}^d_{,a}\,\bar{q}^e_{,b}\ ,\ \bar{q}^c_{,d} := \partial\bar{q}^c/\partial q^d \tag{2.1–3a}$$

are of fundamental importance. For transformations between Cartesian and curvilinear coordinates $q^a \Leftrightarrow x^i \equiv x_{i.}$ follow from this the differential equations or

derivation equations

$$x^k_{,ab} = \Gamma^d_{ab}\, x^k_{,d}\,, \qquad q^c_{,ij} = -\Gamma^c_{de}\, q^d_{,i}\, q^e_{,j} \tag{2.1–3b}$$

with the covariant and contravariant

basis vectors

$$x^k_{,d} = x_{k,d} := \partial x_{k.}/\partial q^d =: c_{k.d}\,,\ q^d_{,i} := \partial q^d/\partial x_{i.} =: c^d_{i.}\,; \tag{2.1–3c}$$

$$\Gamma^d_{ab} = \textit{Christoffel symbols of the metric of the } q^a\,.$$

In the surface geometry the

Gaussian surface representation in Cartesian coordinates

$$x_{i.} = x_{i.}(u^\alpha)\,, \tag{2.1–4a}$$

is used in particular, among others, for the special case

$$u^\alpha =: x_{\alpha.}\,, \qquad \alpha \in \{1,2\}\ :$$

$$x_{i.}(x_{\alpha.}) = [\,x_{1.}\,, x_{2.}\,, x_{3.}(x_{\alpha.})\,]\,. \tag{2.1–4b}$$

Analogously, in curvilinear coordinates:

$$q^a = q^a(u^\alpha) \tag{2.1–5a}$$

with the special case

$$q^\alpha =: u^\alpha\,, \qquad \alpha \in \{1,\,2\}\ :$$

$$q^a(u^\alpha) = [\,u^1,\, u^2,\, q^3(u^\alpha)\,]\,. \tag{2.1–5b}$$

2.1.2 Surface Normal Coordinates

Surface normal coordinates are introduced according to Heitz (1988). For primary space and surface representation

external surface normal coordinates:

$$q^a = (u^1, u^2, q^3 \equiv H) = (u^\alpha, H), \tag{2.1–6a}$$

$$u^\alpha = \textit{surface coordinates of the reference surface} \tag{2.1–6b}$$

are used with

$$\begin{aligned} &\textit{metric tensor } g_{ab}, \\ &\textit{basis vectors } c_{i.a} := \partial x_{i.}/\partial q^a, \quad c^a_{i.} := \partial q^a/\partial x_{i.}. \end{aligned} \tag{2.1–6c}$$

The Gaussian surface representation according to (2.1–5) results in

$$q^a(u^\alpha) = [\, u^1, u^2, H(u^\alpha)\,]. \tag{2.1–6d}$$

An important example with regard to external surface normal coordinates is

geographical ellipsoidal coordinates :

reference surface = ellipsoid of revolution,

$$q^a = (\lambda, \phi, H) = (u^\alpha, H), \tag{2.1–7a}$$

$$\begin{aligned} u^\alpha &= (\lambda, \phi) \\ &= \textit{geographical longitude, latitude of the ellipsoid of revolution}, \\ H &= \textit{ellipsoidal height}. \end{aligned} \tag{2.1–7b}$$

For the representation of external surface deformations the

internal surface normal coordinates (*Fig. 2.1*)

$$p^a = (p^1, p^2, p^3 \equiv h) = (p^\alpha, h), \tag{2.1–8a}$$

$$\begin{aligned} p^\alpha &= \textit{surface coordinates of the reference surface} \\ &= \textit{surface considered in the undeformed state} \end{aligned} \tag{2.1–8b}$$

are used with

$$\begin{aligned} &\textit{metric tensor } f_{ab}, \\ &\textit{basis vectors } b_{i.a} := \partial x_{i.}/\partial p^a, \quad b^a_{i.} := \partial p^a/\partial x_{i.}. \end{aligned} \tag{2.1–8c}$$

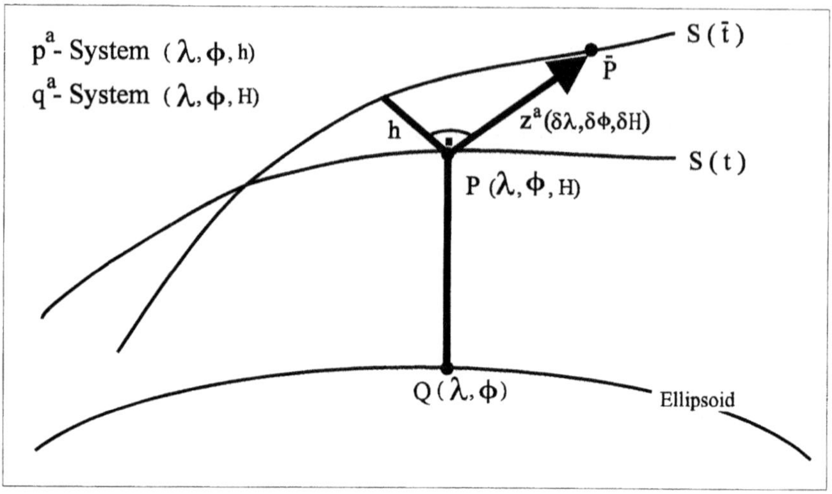

Fig. 2.1. Relations between the internal surface normal coordinates and external surface normal coordinates

For the investigations of surface deformations, external and internal surface normal coordinates are generally combined. The primary surface representation is done according to (2.1–6d) and in (2.1–8a) the surface coordinates u^α are adopted:

$$p^a =: (u^\alpha, h) . \qquad (2.1\text{–}8d)$$

2.2 Three–Dimensional Deformations

2.2.1 Preliminary Remarks

The deformation theory originates from the representation of the dynamics of deformable media (body). Therefore, it has primarily been developed as a three–dimensional differential geometry of deformable point fields. In this *section 2.2* only the most important general results are described in short, since the main objective of this work is the surface deformations dealt with in the following *sections 2.3* and in *chapter 4* which, when applied to free body surfaces, can be directly derived from observations.

2.2.2 General Fundamentals

2.2.2.1 Isoparametric Representation

A point field is considered

$$\{P\} \in \textit{body volume } V \tag{2.2–1a}$$

at two points of time

$$t \textit{ and } \bar{t} \tag{2.2–1b}$$

in the
isoparametric representation

$$q^a = \textit{coordinates of } P \textit{ in the points of time } t \textit{ and } \bar{t} \;: \tag{2.2–2a}$$

$$g_{ab} = \textit{metric tensor in the point of time } t \;, \tag{2.2–2b}$$

$$\bar{g}_{ab} = \textit{metric tensor in the point of time } \bar{t}. \tag{2.2–2c}$$

The difference of the square of the linear elements is the primary

deformation measure

$$d\bar{s}^2 - ds^2 = (\bar{g}_{ab} - g_{ab})\, dq^a\, dq^b = 2\, D_{ab}\, dq^a\, dq^b \tag{2.2–3a}$$

with the
deformation tensor:

$$D_{ab} = (\bar{g}_{ab} - g_{ab})/2 = D_{ba}\,. \tag{2.2–3b}$$

2.2.2.2 Lagrangian Representation

The Lagrangian representation is achieved by introducing the position vectors for the isoparametric representation (2.2–2):

$$\left.\begin{aligned} x_{i.}(q^a) &= \textit{Cartesian coordinates of the } P(t) \\ &\quad\; \textit{in the system } S \;, \\ \bar{x}_{i.}(q^a) &= \textit{Cartesian coordinates of the } P(\bar{t}) \\ &\quad\; \textit{in the system } \bar{S} \equiv S, \end{aligned}\right. \tag{2.2–4a}$$

by means of which the

displacement vectors in $S \equiv \bar{S}$ are defined as follows:

$$z_{i.}(q^a) := \bar{x}_{i.}(q^a) - x_{i.}(q^a). \tag{2.2–4b}$$

For the

metric tensors

$$g_{ab} = x_{i,a}\, x_{i,b}\,, \qquad \bar{g}_{ab} = \bar{x}_{i,a}\, \bar{x}_{i,b} \tag{2.2–5a}$$

the relation

$$\bar{g}_{ab} = g_{ab} + x_{i,a}\, z_{i,b} + x_{i,b}\, z_{i,a} + z_{i,a}\, z_{i,b}\,, \tag{2.2–5b}$$

is obtained, through which the

deformation tensor to

$$\begin{aligned} D_{ab} &= (x_{i,a}\, z_{i,b} + x_{i,b}\, z_{i,a} + z_{i,a}\, z_{i,b})/2 \qquad (2.2\text{–}5\text{c}) \\ &= \varepsilon_{ab} + z_{i,a}\, z_{i,b}/2\,, \end{aligned}$$

$$\varepsilon_{ab} := (x_{i,a}\, z_{i,b} + x_{i,b}\, z_{i,a})/2\,, \tag{2.2–5d}$$

results. Of special importance from a practical point of view is the case of

small displacements,

$$|z_{i,a}| \ll |x_{i,a}|\,, \tag{2.2–6a}$$

to which applies sufficiently precisely the

linear deformation tensor

$$D_{ab} \approx \varepsilon_{ab} = (x_{i,a}\, z_{i,b} + x_{i,b}\, z_{i,a})/2\,. \tag{2.2–6b}$$

By transforming the deformation tensor (2.2–5c) into Cartesian coordinates, one obtains with the introduction of the

displacement tensor

$$V_{ij.} := \partial_{j.} z_{i.} \tag{2.2–7a}$$

the

Cartesian deformation tensor

$$D_{ij.} := \varepsilon_{ij.} + V_{ki.}V_{kj.}/2 \; = \; D_{ji.} \tag{2.2–7b}$$

with

$$\varepsilon_{ij.} = (V_{ij.} + V_{ji.})/2 \; = \; (\partial_{i.}z_{j.} + \partial_{j.}z_{i.})/2\,. \tag{2.2–7c}$$

With (2.2–7b, c) and with the

tensor of rotation

$$\xi_{ij.} = (V_{ji.} - V_{ij.})/2 \; = \; (\partial_{i.}z_{j.} - \partial_{j.}z_{i.})/2\,, \tag{2.2–7d}$$

the tensor of displacement to

$$V_{ij.} = \varepsilon_{ij.} - \xi_{ij.} \tag{2.2–7e}$$

results. The deformation measure (2.2–3a) is given the form

$$d\bar{s}^2 - ds^2 \;\; = \;\; 2D_{ij.}\,dx_{i.}dx_{j.} \tag{2.2–7f}$$

$$\approx \;\; 2\varepsilon_{ij.}\,dx_{i.}dx_{j.} \quad \textit{for small displacements.} \tag{2.2–7g}$$

Special representations of rotations are

$$\xi_{ij.} = \varepsilon_{ijk.}\tilde{d}_{k.} \qquad\qquad \Rightarrow \;\; \tilde{d}_{i.} = \varepsilon_{ijk.}\,\partial_{j.}z_{k.}/2\;, \tag{2.2–8a}$$

and with

$$v_{i.} := dz_{i.}/dt\;, \qquad\qquad \tilde{w}_{i.} := \tilde{d}_{i.}/dt\;\;:$$

$$\tilde{w}_{i.} \;\; = \;\; \varepsilon_{ijk.}\,\partial_{j.}v_{k.} \tag{2.2–8b}$$

$$= \;\; \textit{vorticity, eddy (vortex), rotation of } v_{k.}\,.$$

2.2.3 Special Deformation Measures

The two most important deformation measures are

$$q \;:=\; (d\bar{s} - ds)/ds \;=\; \textit{linear elongation}\;, \tag{2.2–9a}$$

$$\bar{q} \;:=\; (d\bar{V} - dV)/dV \;=\; \textit{volume dilatation}\;. \tag{2.2–9b}$$

From (2.2–3a) it follows directly for the

Linear elongation
in the direction of $r^a := dq^a/ds$ bzw. $r_{i.} := dx_{i.}/ds$:

$$q = (1 + 2\,D_{ab}\,r^a r^b)^{1/2} - 1 = (1 + 2\,D_{ij.}\,r_{i.} r_{j.})^{1/2} - 1\;, \tag{2.2–10a}$$

and in the case of small displacements (2.2–6) these formulas simplify themselves as follows:

$$q = \varepsilon_{ab}\,r^a\,r^b = \varepsilon_{ij.}\,r_{i.}\,r_{j.}\;. \tag{2.2–10b}$$

For the volume dilatation (2.2–9b) the representation of Duschek and Hochrainer (1965–1968) §43 is adopted. In this case, the *'first, second, and third scalar of the deformation tensor'* are

$$D^{(1)} \;=\; \varepsilon_{ijk.}\varepsilon_{pqr.}\delta_{ip.}\delta_{jq.}D_{kr.}/2 = D_{ii.}\;, \tag{2.2–11a}$$

$$D^{(2)} \;=\; \varepsilon_{ijk.}\varepsilon_{pqr.}\delta_{ip.}D_{jq.}D_{kr.}/2\;, \tag{2.2–11b}$$

$$D^{(3)} \;=\; \varepsilon_{ijk.}\varepsilon_{pqr.}D_{ip.}D_{jq.}D_{kr.}/6\;, \tag{2.2–11c}$$

where

$$\begin{aligned} \varepsilon_{ijk} \;=\; \varepsilon^{ijk} \;&=\; +1 \quad \textit{if} \;\; i,j,k \;\; \textit{are cyclic},\\ &=\; -1 \quad \textit{if} \;\; i,j,k \;\; \textit{are anticyclic},\\ &=\; \;\;0 \quad \textit{if} \;\; i,j,k \;\; \textit{are acyclic}, \end{aligned} \tag{2.2–11d}$$

are the permutation symbols for the Cartesian coordinates. For the permutation symbols of the ellipsoidal coordinates see (2.3–27d).

With (2.2–11a–c) one obtains for the

volume dilatation (–depression):

$$\bar{q} = (1 + 2\,D^{(1)} + 4\,D^{(1)} + 8\,D^{(3)})^{1/2} - 1\;. \tag{2.2–12a}$$

In the case of small displacements (2.2–6), (2.2–11a–c) simplify themselves as follows:

$$D^{(1)} = \varepsilon_{ii.} \,, \qquad D^{(2)} \approx 0 \,, \qquad D^{(3)} \approx 0 \,,$$

by which (2.2–12a) goes over into

$$\bar{q} = \varepsilon_{ii.} = g^{ab}\, \varepsilon_{ab} \,. \qquad (2.2\text{–}12b)$$

2.3 Surface Deformations

2.3.1 Preliminary Remarks

If one considers only the internal Riemannian geometry of surfaces and the deformations referred to it, then the pertinent theory is formally, to a large extent, identical with the three–dimensional deformation theory dealt with in *section 2.2.* This is directly apparent from the following *sections 2.3.2.1 and 2.3.2.2.* However, as regards the applications, the embedding of the surfaces into the three–dimensional Euclidean space is of interest, which means the change resulting from surface deformations of the external geometry, the fundamentals of which are treated in *section 2.3.3.*

2.3.2 Internal Surface Deformations

2.3.2.1 Isoparametric Representation

Here, a two–dimensional point field

$$\{P\} \in \textit{surface } F \qquad (2.3\text{–}1a)$$

is at two points of time

$$t \;\; \textit{and} \;\; \bar{t} \qquad (2.3\text{–}1b)$$

in the **isoparametric representation**

$$u^{\alpha} \;=\; \textit{coordinates of } P \;\; \textit{in the moments } t \text{ and } \bar{t} \,, \qquad (2.3\text{–}2a)$$

$$f_{\alpha\beta} \;=\; \textit{metric tensor in the moment } t \,, \qquad (2.3\text{–}2b)$$

$$\bar{f}_{\alpha\beta} \;=\; \textit{metric tensor in the moment } \bar{t} \,, \qquad (2.3\text{–}2c)$$

taken into account analogously to (2.2–1). The difference of the squares of the linear elements is the primary

deformation measure

$$d\bar{s}^2 - ds^2 = (\bar{f}_{\alpha\beta} - f_{\alpha\beta})\, du^\alpha\, du^\beta = 2\, D_{\alpha\beta}\, du^\alpha\, du^\beta \qquad (2.3\text{–}3a)$$

with the

surface deformation tensor:

$$D_{\alpha\beta} = (\bar{f}_{\alpha\beta} - f_{\alpha\beta})/2 = D_{\beta\alpha}\,. \qquad (2.3\text{–}3b)$$

2.3.2.2 Computations on the Basis of the Gaussian Surface Representation

The position vectors

$$\begin{aligned} x_{i.}(u^\alpha) &= \textit{Cartesian coordinates of the } P(t) \in F \\ &\quad\ \textit{in the system } S\,, \\ \bar{x}_{i.}(u^\alpha) &= \textit{Cartesian coordinates of the } P(\bar{t}) \in \bar{F} \\ &\quad\ \textit{in the system } \bar{S} \equiv S \end{aligned} \qquad (2.3\text{–}4a)$$

are introduced into the isoparametric surface representation (2.3–2), by means of which the

displacement vectors in $S \equiv \bar{S}$ are formed:

$$z_{i.}(u^\alpha) := \bar{x}_{i.}(u^\alpha) - x_{i.}(u^\alpha)\,; \qquad (2.3\text{–}4b)$$

For the

metric tensors

$$f_{\alpha\beta} = x_{i,\alpha}\, x_{i,\beta}\,, \qquad \bar{f}_{\alpha\beta} = \bar{x}_{i,\alpha}\, \bar{x}_{i,\beta} \qquad (2.3\text{–}5a)$$

the relation

$$\bar{f}_{\alpha\beta} = f_{\alpha\beta} + x_{i,\alpha}\, z_{i,\beta} + x_{i,\beta}\, z_{i,\alpha} + z_{i,\alpha}\, z_{i,\beta}\,, \qquad (2.3\text{–}5b)$$

is obtained, which leads to the

surface deformation tensor

$$D_{\alpha\beta} = (x_{i,\alpha}\, z_{i,\beta} + x_{i,\beta}\, z_{i,\alpha} + z_{i,\alpha}\, z_{i,\beta})/2 \qquad (2.3\text{–}5c)$$

$$= \varepsilon_{\alpha\beta} + z_{i,\alpha}\, z_{i,\beta}/2 \; ,$$

$$\varepsilon_{\alpha\beta} = (x_{i,\alpha}\, z_{i,\beta} + x_{i,\beta}\, z_{i,\alpha})/2 \, . \qquad (2.3\text{–}5d)$$

To

small displacements

$$|z_{i,\alpha}| \ll |x_{i,\alpha}| \; , \qquad (2.3\text{–}6a)$$

the **linear surface deformation tensor**

applies sufficiently precisely

$$D_{\alpha\beta} \approx \varepsilon_{\alpha\beta} = (x_{i,\alpha}\, z_{i,\beta} + x_{i,\beta}\, z_{i,\alpha})/2 \; , \qquad (2.3\text{–}6b)$$

and according to (2.2–10b, 12b) the **internal surface–deformation–measures** for small displacements result as follows:

Linear surface elongation q and the **scale of distortion m** in the direction of $r^\alpha := du^\alpha/ds$:

$$q := (d\bar{s} - ds)/ds = \varepsilon_{\alpha\beta}\, r^\alpha\, r^\beta \; , m = 1 + q \; , \qquad (2.3\text{–}7a)$$

Surface dilatation (–depression):

$$\bar{q} = f^{\alpha\beta}\, \varepsilon_{\alpha\beta} \, . \qquad (2.3\text{–}7b)$$

For the

main deformation directions orthogonal to each other

$$\nu_{(n)} := (du^2/du^1)_n \; , \qquad n \in \{1, 2\} \; , \qquad (2.3\text{–}7c)$$

which are identical with the directions of the principal axes of the linear deformation tensor $\varepsilon_{\alpha\beta}$, one obtains starting from (2.3–7a) as the conditional equation the quadratic equation

$$\begin{aligned} &a_0\, \nu^2 + a_1\, \nu + a_2 = 0 \; , && a_0 = \varepsilon_{12}\, f_{22} - \varepsilon_{22}\, f_{12} \; , \\ &a_1 = \varepsilon_{11}\, f_{22} - \varepsilon_{22}\, f_{11} \; , && a_2 = \varepsilon_{11}\, f_{12} - \varepsilon_{12}\, f_{11} \end{aligned} \qquad (2.3\text{–}7d)$$

The two solutions of the quadratic equation generally read

$$\left. \begin{matrix} \nu_1 \\ \nu_2 \end{matrix} \right\} = [1/(2a_0)]\,[\, -a_1 \pm (a_1^2 - 4\, a_0\, a_2)^{1/2}\,] \, . \qquad (2.3\text{–}7e)$$

For the computation of the quadratic equation (2.3–7e) the special cases

$$\begin{aligned} &|a_0\, a_2| \ll a_1^2 \ : \\ &\nu_1 \to 0 \ , \qquad\qquad \nu_2 \to -a_1/a_0 \ , \\ &|a_2| \ll |a_1| \ \ and \ \ |a_0| \ll |a_1| \ : \\ &\nu_1 \to 0 \ , \qquad\qquad \nu_2 \to \infty \end{aligned} \tag{2.3–7f}$$

are of interest. Another special case for

$$\varepsilon_{12} \to 0 \qquad and \qquad f_{12} \to 0$$

is given if also locally the main deformation directions coincide with orthogonal coordinate lines. To the main deformation directions (2.3–7e, f) belong the direction vectors

$$\begin{aligned} &r^\alpha_{(n)} = (1\,,\ \nu_{(n)})^\alpha / |(1\,,\ \nu_{(n)})^\beta| \ , \\ &|(1\,,\ \nu_{(n)})^\beta|^2 = f_{\alpha\beta}\,(1\,,\ \nu_{(n)})^\alpha\,(1\,,\ \nu_{(n)})^\beta \ . \end{aligned} \tag{2.3–7g}$$

Owing to (2.3–7g) one obtains the components of the direction vectors

$$r^1_{(1)} = 1\,/\,(1+\nu^2_{(1)})^{1/2} \ , \qquad r^1_{(2)} = 1\,/\,(1+\nu^2_{(2)})^{1/2} \tag{2.3–7h}$$

and

$$r^2_{(1)} = \nu_{(1)}\,/\,(1+\nu^2_{(1)})^{1/2} \ , \ r^2_{(2)} = \nu_{(2)}\,/\,(1+\nu^2_{(2)})^{1/2}. \tag{2.3–7i}$$

If these are inserted into (2.3–7a) the

extreme values of the linear elongations and **distortion scales**

$$q_{(n)} = \varepsilon_{\alpha\beta}\, r^\alpha_{(n)}\, r^\beta_{(n)} \ , \qquad m_{(n)} = 1 + q_{(n)} \ , \quad n \in \{1,\, 2\} \tag{2.3–7j}$$

and the distortion vectors

$$m^\alpha_{(n)} = m_{(n)}\, r^\alpha_{(n)} \ , \tag{2.3–7k}$$

result.

2.3.2.3 Computations in External Surface Normal Coordinates

In (2.3–4a) the position vectors are replaced by external surface normal coordinates (2.1–6), which leads to the following isoparametric surface representations for the points at time t and $\bar{t}$

$$q^a(u^\alpha) = [u^1, u^2, H(u^\alpha)]^a ,$$

$$\bar{q}^a(u^\alpha) = [\bar{u}^1(u^\alpha), \bar{u}^2(u^\alpha), \bar{H}(u^\alpha)]^a . \tag{2.3–8a}$$

To the displacement vectors (2.3–4b) correspond here the

displacement coordinates

$$z^a(u^\alpha) := \bar{q}^a(u^\alpha) - q^a(u^\alpha) . \tag{2.3–8b}$$

Only the case of small displacements (2.3–6a) is considered, for which with (2.3–8a)

$$|z^a_{,\alpha}| \ll |q^a_{,\alpha}| . \tag{2.3–9a}$$

With this accuracy one obtains in the q^a–system :

$$z^a = c^a_{i.} z_{i.} ,$$

$$z_{i.} = (\partial x_{i.}/\partial q^a)\, \delta q^a = c_{i.a}\, z^a , \tag{2.3–9b}$$

$$\bar{q}^a = q^a + z^a ,$$

and it is

$$x_{i,\alpha}\, z_{i,\beta} = q^c_{,\alpha}\,(z^d_{,\beta} + \Gamma^d_{ef}\, q^e_{,\beta}\, z^f)\, g_{cd} . \tag{2.3–9c}$$

According to (2.3–5a) it follows for the

metric tensors

$$f_{\alpha\beta} = q^c_{,\alpha}\, q^d_{,\beta}\, g_{cd} , \qquad \bar{f}_{\alpha\beta} = \bar{q}^c_{,\alpha}\, \bar{q}^d_{,\beta}\, \bar{g}_{cd} \tag{2.3–10a}$$

of F and $\bar{F}$ the relation

$$\bar{f}_{\alpha\beta} = f_{\alpha\beta} + [q^c_{,\alpha}\,(z^d_{,\beta} + \Gamma^d_{ef}\, q^e_{,\beta}\, z^f) + q^c_{,\beta}\,(z^d_{,\alpha} + \Gamma^d_{ef}\, q^e_{,\alpha}\, z^f) + \ldots]\, g_{cd} , \tag{2.3–10b}$$

so that the

linear surface deformation tensor to

$$\varepsilon_{\alpha\beta} = [\, q^c_{,\alpha} \, (z^d_{,\beta} + \Gamma^d_{ef} \, q^e_{,\beta} \, z^f) + q^c_{,\beta} \, (z^d_{,\alpha} + \Gamma^d_{ef} \, q^e_{,\alpha} \, z^f) \,] \, g_{cd}/2 \qquad (2.3\text{–}10c)$$

results. The first derivations of the q^c according to the surface coordinates result on the basis of (2.3–8a) to

$$q^c_{,\alpha} = (\, \delta^1_\alpha \, , \, \delta^2_\alpha \, , \, H_{,\alpha} \,)^c \, . \qquad (2.3\text{–}10d)$$

For the **internal surface deformation measures** here also apply (2.3–7a–k).

2.3.2.4 Computations in Ellipsoidal Coordinates

For the surface coordinates (2.3–8a) geographical coordinates are selected

$$u^\alpha = (\lambda, \, \phi) \, , \qquad \alpha \in \{1,2\} \, , \qquad (2.3\text{–}11a)$$

so that the surface representation for successive points in time t and $\bar{t}$ with (2.1–6) to

$$\begin{aligned} q^a(u^\alpha) &= [\lambda, \, \phi, \, H(u^\alpha)]^a \\ \bar{q}^a(u^\alpha) &= [\bar{\lambda}(u^\alpha), \, \bar{\phi}(u^\alpha), \, \bar{H}(u^\alpha)]^a \end{aligned} \qquad (2.3\text{–}11b)$$

and the displacement coordinates (2.3–8b) to

$$\delta q^a(u^\alpha) = [\, \delta\lambda(u^\alpha), \, \delta\phi(u^\alpha), \, \delta H(u^\alpha) \,]^a \qquad (2.3\text{–}11c)$$

result. The metric of the geographical ellipsoidal coordinates is determined by

$$g_{ab} = (R_1 + H)^2 \cos^2\phi \; \delta^1_a \delta^1_b + (R_2 + H)^2 \; \delta^2_a \delta^2_b + \delta^3_a \delta^3_b \qquad (2.3\text{–}12a)$$

with

$R_1 = c/V$, $R_2 = c/V^3$, $c = a^2/b$,

a = *semi major axis of the ellipsoid of revolution,*

b = *semi minor axis of the ellipsoid of revolution,* and

$V = (1 + e''^2 \cos^2\phi)^{1/2}$, $e''^2 = (a^2 - b^2) \, / \, b^2$.

The first partial derivatives of the metric tensor (2.3–12a) resulting from taking λ, ϕ and H read

$$
\begin{aligned}
g_{11,1} &= 0 \\
g_{11,2} &= -(R_1 + H)(R_2 + H)\sin 2\phi \\
g_{11,3} &= 2(R_1 + H)\cos^2\phi \\
g_{22,1} &= 0 \\
g_{22,2} &= 3R_2(R_2 + H)(e''/V)^2 \sin 2\phi \\
g_{22,3} &= 2(R_2 + H) \\
g_{33,1} &= g_{33,2} = g_{33,3} = 0
\end{aligned}
\qquad (2.3\text{–}12b)
$$

with
e''^2= *2nd numerical eccentricity of the ellipsoid of revolution (2.3–12a).*

From (2.3–12) one obtains the Christoffel symbols of the 2nd kind needed in (2.3–9c)

$$\Gamma^c_{ab} = g^{cd}\,(g_{ad,b} + g_{bd,a} - g_{ab,d})\,/\,2, \qquad (2.3\text{–}13a)$$

as follows:

$$
\begin{aligned}
&\Gamma^1_{12} = -(R_2 + H)\tan\phi\,/\,(R_1 + H)\ , \\
&\Gamma^1_{13} = 1\,/\,(R_1 + H)\ , \\
&\Gamma^1_{21} = \Gamma^1_{12}\ , \\
&\Gamma^1_{31} = \Gamma^1_{13}\ , \\
&\Gamma^2_{11} = (R_1 + H)\sin 2\phi\,/\,2(R_2 + H)\ , \\
&\Gamma^2_{22} = 3R_2\,(e''/V)^2\sin 2\phi\,/\,2(R_2 + H)\ , \\
&\Gamma^2_{23} = 1\,/\,(R_2 + H)\ , \\
&\Gamma^2_{32} = \Gamma^2_{23}\ , \\
&\Gamma^3_{11} = -(R_1 + H)\cos^2\phi\ , \\
&\Gamma^3_{22} = -(R_2 + H)\ .
\end{aligned}
\qquad (2.3\text{–}13b)
$$

The other components of the Christoffel symbols of the 2nd kind, which are not included in (2.3–13b) equal to zero. For (2.3–9c) it follows from this, using the assumptions (2.3–9a, b) with the displacement coordinates (2.3–11c), in geographical ellipsoidal coordinates

$$
\begin{aligned}
x_{i,1}z_{i,1} &= \left[z^1_{,1} + \left(\Gamma^1_{1a} + \Gamma^1_{3a}H_{,1}\right) z^a\right] g_{11} \\
&\quad + \left[z^3_{,1} + \left(\Gamma^3_{1a} + \Gamma^3_{3a}H_{,1}\right) z^a\right] H_{,1} \, , \\
x_{i,1}z_{i,2} &= \left[z^1_{,2} + \left(\Gamma^1_{2a} + \Gamma^1_{3a}H_{,2}\right) z^a\right] g_{11} \\
&\quad + \left[z^3_{,2} + \left(\Gamma^3_{2a} + \Gamma^3_{3a}H_{,2}\right) z^a\right] H_{,1} \, , \qquad (2.3\text{–}14) \\
x_{i,2}z_{i,1} &= \left[z^2_{,1} + \left(\Gamma^2_{1a} + \Gamma^2_{3a}H_{,1}\right) z^a\right] g_{22} \\
&\quad + \left[z^3_{,1} + \left(\Gamma^3_{1a} + \Gamma^3_{3a}H_{,1}\right) z^a\right] H_{,2} \, , \\
x_{i,2}z_{i,2} &= \left[z^2_{,2} + \left(\Gamma^2_{2a} + \Gamma^2_{3a}H_{,2}\right) z^a\right] g_{22} \\
&\quad + \left[z^3_{,2} + \left(\Gamma^3_{2a} + \Gamma^3_{3a}H_{,2}\right) z^a\right] H_{,2} \, .
\end{aligned}
$$

With the aid of the law of transformation of the metric tensor

$$f_{\alpha\beta} = q^c_{,\alpha} q^d_{,\beta} \, g_{cd} \qquad (2.3\text{–}15)$$

one obtains the metric tensor $f_{\alpha\beta}$ in the first point of time t in the isoparametric surface representation (2.3–2)

$$
\begin{aligned}
f_{\alpha\beta} &= [(R_1 + H)^2 \cos^2\phi + (H_{,1})^2]\delta^1_\alpha \, \delta^1_\beta \qquad (2.3\text{–}16) \\
&\quad + (\delta^1_\alpha \, \delta^2_\beta + \delta^2_\alpha \, \delta^1_\beta) H_{,1} \, H_{,2} + [(R_2 + H)^2 + (H_{,2})^2] \, \delta^2_\alpha \, \delta^2_\beta \, .
\end{aligned}
$$

Thus, all relevant quantities have been determined to be able to compute the metric tensor $\bar{f}_{\alpha\beta}$, (2.3–10a) for the second point of time $\bar{t}$, and the linear surface deformation tensor $\epsilon_{\alpha\beta}$, (2.3–10c). The **linear surface elongation** and the **surface dilatation** (–depression) result according to (2.3–7).

2.3.3 External Surface Deformations

2.3.3.1 Computations on the Basis of the Gaussian Surface Representation

External surface deformations are primarily determined by the **changes of the first and second fundamental tensor**

$$\bar{f}_{\alpha\beta} - f_{\alpha\beta} \, , \qquad \bar{L}_{\alpha\beta} - L_{\alpha\beta} \qquad (2.3\text{–}17a)$$

For small displacements (2.3–6) the

change of the first fundamental tensor can be computed according to (2.3–5) as follows:

$$\delta f_{\alpha\beta} = \bar{f}_{\alpha\beta} - f_{\alpha\beta} = x_{i,\alpha}\, z_{i,\beta} + x_{i,\beta}\, z_{i,\alpha} + \ldots = 2\,\varepsilon_{\alpha\beta} \,, \qquad (2.3\text{–}17b)$$

$$\begin{aligned} \delta f = \bar{f} - f &= f_{11}\,\delta f_{22} + f_{22}\,\delta f_{11} - 2\, f_{12}\,\delta f_{12} + \ldots \qquad (2.3\text{–}17c) \\ &= 2\,(\, f_{11}\,\varepsilon_{22} + f_{22}\,\varepsilon_{11} - 2\, f_{12}\,\varepsilon_{12} + \ldots) \,. \end{aligned}$$

With this result for the second fundamental tensor of F and $\bar{F}$

$$\begin{aligned} L_{\alpha\beta} &= n_{i.}\, x_{i,\alpha\beta} = \varepsilon_{ijk.}\, x_{j,1}\, x_{k,2}\, x_{i,\alpha\beta}\, f^{-1/2} \,, \\ \bar{L}_{\alpha\beta} &= \bar{n}_{i.}\, \bar{x}_{i,\alpha\beta} = \varepsilon_{ijk.}\, \bar{x}_{j,1}\, \bar{x}_{k,2}\, \bar{x}_{i,\alpha\beta}\, \bar{f}^{-1/2} \end{aligned} \qquad (2.3\text{–}18a)$$

and for the

change of the second fundamental tensor in linearized form

$$\begin{aligned} \delta L_{\alpha\beta} &= \bar{L}_{\alpha\beta} - L_{\alpha\beta} \qquad (2.3\text{–}18b) \\ &= \varepsilon_{ijk.}\, \big[x_{j,1}\, x_{k,2}\, z_{i,\alpha\beta} \\ &\qquad + (x_{j,1}\, z_{k,2} + x_{k,2}\, z_{j,1})\, x_{i,\alpha\beta}\big] \,/\, f^{1/2} - L_{\alpha\beta}\,\delta f/(2f) \,, \end{aligned}$$

$$\begin{aligned} \delta L &= \bar{L} - L \qquad (2.3\text{–}18c) \\ &= L_{11}\,\delta L_{22} + L_{22}\,\delta L_{11} - 2\, L_{12}\,\delta L_{12} + \delta L_{11}\,\delta L_{22} + \delta L_{12}^2 \,. \end{aligned}$$

For δL no linearized formula according to (2.3–17c) can be used because the $\delta L_{\alpha\beta}$ may be of the same order of magnitude as the $L_{\alpha\beta}$. This is particularly important as regards the computation of the change of the Gaussian curvature (2.3–21d).

On the other hand the external surface geometry, which is more graphic, is determined by the

normal curvatures
in the directions $r^\alpha := du^\alpha/ds$:

$$\kappa_N = 1/R_N = L_{\alpha\beta}\, r^\alpha\, r^\beta \,, \qquad (2.3\text{–}19a)$$

the

mean curvature

$$H = (\kappa_1 + \kappa_2)/2 = f^{\alpha\beta}\, L_{\alpha\beta}/2 \tag{2.3–19b}$$

and the

Gaussian curvature

$$K = (\kappa_1\, \kappa_2)^{1/2} = (L/f)^{1/2}, \tag{2.3–19c}$$

defined here as geometrical mean, by

$$f = |f_{\alpha\beta}| = f_{11}\, f_{22} - f_{12}^2\,, \quad L = |L_{\alpha\beta}| = L_{11}\, L_{22} - L_{12}^2\,. \tag{2.3–19d}$$

According to the 'Theorema egregium by Gauss' K can also be represented by itself as a function of the internal geometry (see Heitz 1988, Chapter 2)

$$K^2 = R_{2112}/f\,. \tag{2.3–19e}$$

For the **principal directions of curvature** orthogonal to each other

$$\nu_{(n)} := (du^2/du^1)_n\,, \qquad n \in \{1, 2\}\,, \tag{2.3–20a}$$

which are identical with the directions of the principal axes of the second fundamental tensor $L_{\alpha\beta}$, one obtains starting from (2.3–19a) as conditional equation the quadratic equation

$$\begin{aligned} &b_0\,\nu^2 + b_1\,\nu + b_2 = 0\,, &\quad b_0 &= L_{12}\, f_{22} - L_{22}\, f_{12}\,, \\ &b_1 = L_{11}\, f_{22} - L_{22}\, f_{11}\,, &\quad b_2 &= L_{11}\, f_{12} - L_{12}\, f_{11} \end{aligned} \tag{2.3–20b}$$

(see (2.3–7) and Heitz 1988, Chapter 2). The two solutions of the quadratic equations generally read

$$\left.\begin{matrix} \nu_{(1)} \\ \nu_{(2)} \end{matrix}\right\} = [1/(2b_0)]\,[-b_1 \pm (b_1^2 - 4\, b_0\, b_2)^{1/2}]. \tag{2.3–20c}$$

In this context the special cases

$$\begin{aligned} &|b_0\, b_2| \ll b_1^2 : \\ &\nu_{(1)} \to 0\,, \qquad\qquad \nu_{(2)} \to -b_1/b_0\,, \\ &|b_2| \ll |b_1| \;\; \mathit{and} \;\; |b_0| \ll |b_1| : \\ &\nu_{(1)} \to 0\,, \qquad\qquad \nu_{(2)} \to \infty \end{aligned} \tag{2.3–20d}$$

are of importance. The second special case is given if

$$L_{12} \to 0 \qquad \text{and} \qquad f_{12} \to 0,$$

i.e. if the principal directions of curvature coincide locally with the orthogonal coordinate lines. To the principal directions of curvature (2.3–20c, d) belong the direction vectors

$$\begin{aligned} &r^{\alpha}_{(n)} = (1\,,\, \nu_{(n)})^{\alpha} / |(1\,,\, \nu_{(n)})^{\beta}|\;, \\ &|(1\,,\, \nu_{(n)})^{\beta}|^2 = f_{\alpha\beta}\,(1\,,\, \nu_{(n)})^{\alpha}\,(1\,,\, \nu_{(n)})^{\beta}\;. \end{aligned} \tag{2.3–20e}$$

According to (2.3–20e) the components of the direction vectors

$$r^1_{(1)} = 1\,/\,(1+\nu^2_{(1)})^{1/2}\;, \qquad r^1_{(2)} = 1\,/\,(1+\nu^2_{(2)})^{1/2} \tag{2.3–20f}$$

and

$$r^2_{(1)} = \nu_{(1)}\,/\,(1+\nu^2_{(1)})^{1/2}\;, \quad r^2_{(2)} = \nu_{(2)}\,/\,(1+\nu^2_{(2)})^{1/2}. \tag{2.3–20g}$$

can be determined. If these are inserted in (2.3–19a) the **principal curvatures**

$$\kappa_{N(n)} = L_{\alpha\beta}\, r^{\alpha}_{(n)}\, r^{\beta}_{(n)}\;, \qquad n \in \{1,\, 2\} \tag{2.3–20h}$$

are obtained, and together with (2.3–20e, f) the principal curvature vectors

$$\kappa^{\alpha}_{N(n)} = \kappa_{N(n)}\, r^{\alpha}_{(n)}\;. \tag{2.3–20i}$$

These results (2.3–20a–i) are formally true also for the deformed surface $\bar{F}$ with

$$\bar{f}_{\alpha\beta} = f_{\alpha\beta} + \delta f_{\alpha\beta}\;, \qquad \bar{L}_{\alpha\beta} = L_{\alpha\beta} + \delta L_{\alpha\beta}\;. \tag{2.3–21a}$$

If the $\bar{\nu}_{(n)}$ and $\bar{r}^{\alpha}_{(n)}$ have been computed analogously to (2.3–20a–i), one obtains starting from (2.3–20h) the

changes of the principal curvatures

$$\begin{aligned} \delta\kappa_{N(n)} &= \bar{\kappa}_{N(n)} - \kappa_{N(n)} = \bar{L}_{\alpha\beta}\, \bar{r}^{\alpha}_{(n)}\, \bar{r}^{\beta}_{(n)} - L_{\alpha\beta}\, r^{\alpha}_{(n)}\, r^{\beta}_{(n)} \\ &= \delta L_{\alpha\beta}\, r^{\alpha}_{(n)}\, r^{\beta}_{(n)} + \bar{L}_{\alpha\beta}\,(2\, r^{\alpha}_{(n)} + \delta r^{\alpha}_{(n)})\, \delta r^{\beta}_{(n)}\;, \\ \delta r^{\alpha}_{(n)} &= \bar{r}^{\alpha}_{(n)} - r^{\alpha}_{(n)}\;. \end{aligned} \tag{2.3–21b}$$

These changes of the principal curvatures are generally not identical with the extreme values of the changes of curvature. As further external surface deformation measures, the

changes of the mean curvature

$$\delta H = \bar{H} - H = (\bar{f}^{\alpha\beta}\,\bar{L}_{\alpha\beta} - f^{\alpha\beta}\,L_{\alpha\beta})/2 \qquad (2.3\text{–}21c)$$

$$= f^{\alpha\beta}\,\delta L_{\alpha\beta} + \bar{L}_{\alpha\beta}\,\delta f^{\alpha\beta}$$

and the

change of the Gaussian curvature

$$\delta K = \bar{K} - K = (\bar{L}/\bar{f})^{1/2} - (L/f)^{1/2} \qquad (2.3\text{–}21d)$$

can be used. It is generally not possible to linearize (2.3–21d) according to (2.3–17, 18) for the reason given on (2.3–18d). Even with small changes of curvature, the changes of the principal curvature directions

$$\bar{\nu}_{(n)} - \nu_{(n)}$$

may turn out to be large.

To represent the external surface deformation, the **rotations of the surface elements** can be referred to in addition to the surface curvatures dealt with before. Their computation is appropriately performed in the internal surface normal coordinates $p^a = (u^\alpha, h)$ according to (2.1–8d), in which with

$$x_{i,a} := \partial x_{i.}/\partial p^a =: b_{i.a}\,, \qquad b^a_{i.} := \partial p^a/\partial x_{i.}\,, \qquad (2.3\text{–}22a)$$

$$z_{i,a} := \partial z_{i.}/\partial p^a$$

holds in particular in the surface points

$$x_{i,3} = b_{i.3} = n_{i.}\,, \qquad z_{i,3} = 0_{i.}\,. \qquad (2.3\text{–}22b)$$

In so far as it serves a better distinction, quantities defined in the system p^a are provided with a tilde ~.

For the displacement tensor one obtains in Cartesian Coordinates (2.1–1)

$$V_{ij.} = \partial_{j.} z_{i.} = b^a_{j.}\, z_{i,a}, \qquad (2.3\text{–}23a)$$

and hence in internal surface normal coordinates

$$\tilde{V}_{ab} = \tilde{z}_{a;b} = x_{i,a}\, x_{j,b}\, V_{ij.} = \delta_b^\beta\, b_{i.a}\, z_{i,\beta} =: \tilde{\varepsilon}_{ab} - \tilde{\xi}_{ab} \ . \tag{2.3–23b}$$

In the system p^a results the

symmetric portion of the displacement tensor to

$$\tilde{\varepsilon}_{ab} = (\tilde{z}_{a;b} + \tilde{z}_{b;a})/2 = (\delta_b^\beta\, b_{i.a}\, z_{i,\beta} + \delta_a^\alpha\, b_{i.b}\, z_{i,\alpha})/2 \ . \tag{2.3–24a}$$

Its 'tangential components' form the

linear surface deformation tensor

$$\tilde{\varepsilon}_{\alpha\beta} = (b_{i.\alpha}\, z_{i,\beta} b_{i.\beta}\, z_{i,\alpha})/2 \equiv \varepsilon_{\alpha\beta} \ . \tag{2.3–24b}$$

The skew–symmetric portion of the displacement tensor is the so–called

tensor of rotation

$$\tilde{\xi}_{ab} = (\tilde{z}_{b;a} \ - \ \tilde{z}_{a;b})/2 = (\delta_b^\beta\, b_{i.b}\, z_{i,\beta} - \delta_a^\alpha\, b_{i.a}\, z_{i,\alpha})/2 \ , \tag{2.3–25a}$$

whose 'tangential components' form the

normal tensor of rotation

$$\tilde{\xi}_{\alpha\beta} = (b_{i.\beta}\, z_{i,\alpha} - b_{i.\alpha}\, z_{i,\beta})/2 \equiv \xi_{\alpha\beta}. \tag{2.3–25b}$$

With (2.3–22b, 25b) one obtains for the

symmetric portion of the displacement tensor the special representation:

$$\tilde{\varepsilon}_{ab} = \tilde{\varepsilon}_{\alpha\beta}\, \delta_a^\alpha \delta_b^\beta + n_{i.}\, z_{i,\alpha}\, (\delta_a^\alpha \delta_b^3 + \delta_b^\alpha \delta_a^3)/2 \ . \tag{2.3–26a}$$

Analogically, the

tensor of rotation (2.3–25a, b) with (2.3–22b) can be described as follows:

$$\tilde{\xi}_{ab} = \tilde{\xi}_{\alpha\beta}\, \delta_a^\alpha \delta_b^\beta \ + \ n_{i.}\, z_{i,\alpha}\, (\delta_a^\alpha \delta_b^3 \ - \ \delta_b^\alpha \delta_a^3)/2 \ . \tag{2.3–26b}$$

The

infinitesimal rotation vector pertinent to the tensor of rotation constitutes in Cartesian coordinates while taking into account (2.3–22b)

$$\begin{aligned} d_{k.} &= (1/2)\,\varepsilon_{ijk.}\,\xi_{ij.} = (1/2)\,\varepsilon_{ijk.}\,b^a_{i.}\,b^b_{j.}\,\tilde{\xi}_{ab} \\ &= (1/2)\,\varepsilon_{ijk.}\,b^a_{i.}\,z_{j,a} = (1/2)\,\varepsilon_{ijk.}\,b^\alpha_{i.}\,z_{j,\alpha}\,. \end{aligned} \tag{2.3–27a}$$

In the internal surface normal coordinates p^a holds

$$\tilde{d}^c = (1/2)\,\tilde{\varepsilon}^{abc}\,\tilde{\xi}_{ab} = (1/2)\,\tilde{\varepsilon}^{abc}\,\tilde{z}_{b;a} = (1/2)\,\tilde{\varepsilon}^{\alpha\,bc}\,b_{i.b}z_{i,\alpha} \tag{2.3–27b}$$

and, separately according to 'planimetric and altimetric components':

$$\begin{aligned} &\tilde{d}^\alpha = \tilde{f}^{-1/2}\,(-z_{i,2}\,,\,z_{i,1})^\alpha\,b_{i.3}/2\;,\;b_{i.3} = n_{i.}\;, \\ &\tilde{d}^3 = \tilde{f}^{-1/2}\,(b_{i.1}\,z_{i,2} - b_{i.2}\,z_{i,1})/2\;, \end{aligned} \tag{2.3–27c}$$

with

the permutation symbols for the ellipsoidal coordinates

$$\begin{aligned} \varepsilon_{def} &= g \;\; if \;\; d,e,f \;\; are\ cyclic, \\ &= -g \;\; if \;\; d,e,f \;\; are\ anticyclic, \\ &= 0 \;\; if \;\; d,e,f \;\; are\ acyclic; \\ \varepsilon^{def} &= 1/g \;\; if \;\; d,e,f \;\; are\ cyclic, \\ &= -1/g \;\; if \;\; d,e,f \;\; are\ anticyclic, \\ &= 0 \;\; if \;\; d,e,f \;\; are\ acyclic, \end{aligned} \tag{2.3–27d}$$

and

$g = |g_{ab}| = g_{11}\,g_{22}$, (see 2.3–12a).

The **partial derivations of the Cartesian rotation vector** (2.3–27a) according to the surface coordinates $p^\beta = q^\beta = u^\beta$ are

$$\begin{aligned} d_{k,\beta} &= (\partial_{l.}d_{k.})\,b_{l.\beta} \\ &= (1/2)\,\varepsilon_{ijk.}[(-\tilde{\Gamma}^\alpha_{\beta\gamma}\,b^\gamma_{i.} + \tilde{L}^\alpha_\beta\,n_{i.})\,z_{j,\alpha} + b^\alpha_{i.}\,z_{j,\alpha\beta}], \end{aligned} \tag{2.3–28a}$$

and for the **covariant derivations of the covariant rotation vector** (2.3–27b, c) one obtains

$$\begin{aligned} \tilde{d}^c_{;\tilde{\beta}} &= b^c_{k.}\,d_{k,\beta} \\ &= (1/2)\,\varepsilon_{ijk.}b^c_{k.}[(-\tilde{\Gamma}^\alpha_{\beta\gamma}\,b^\gamma_{i.} + \tilde{L}^\alpha_\beta\,n_{i.})\,z_{j,\alpha} + b^\alpha_{i.}\,z_{j,\alpha\beta}]\,. \end{aligned} \tag{2.3–28b}$$

(2.3–28a, b) are functions of the second partial derivations of the displacements, and hence of the changes of the normal curvatures due to surface deformation.

2.3.3.2 Computations in External Surface Normal Coordinates

With the Gauss–Weingarten equations for the surface theory

$$x_{i,\alpha} = x_{i,a}\, q^a_{,\alpha} \,, \qquad x_{i,\alpha\beta} = x_{i,d}\,(q^d_{,\alpha\beta} + \Gamma^d_{ab}\, q^a_{,\alpha}\, q^b_{,\beta}) \quad (2.3\text{–}29a)$$

and

$$\varepsilon_{ijk.}\, x_{i,d}\, x_{j,e}\, x_{k,f} = \varepsilon_{def} \,, \quad \varepsilon_{ijk.}\, \bar{x}_{i,d}\, \bar{x}_{j,e}\, \bar{x}_{k,f} = \bar{\varepsilon}_{def} \,, \qquad (2.3\text{–}29b)$$

the representations (2.3–18a) for the second fundamental tensor of F and $\bar{F}$ in the external surface normal coordinates q^a go over into

$$\begin{aligned} L_{\alpha\beta} &= \varepsilon_{def}\, q^e_{,1}\, q^f_{,2}\,(q^d_{,\alpha\beta} + \Gamma^d_{ab}\, q^a_{,\alpha}\, q^b_{,\beta})\, f^{-1/2} \,, \\ \bar{L}_{\alpha\beta} &= \bar{\varepsilon}_{def}\, \bar{q}^e_{,1}\, \bar{q}^f_{,2}\,(\bar{q}^d_{,\alpha\beta} + \bar{\Gamma}^d_{ab}\, \bar{q}^a_{,\alpha}\, \bar{q}^b_{,\beta})\, \bar{f}^{-1/2} \,. \end{aligned} \qquad (2.3\text{–}30a)$$

For the computation of the **change of the second fundamental tensor**

$$\bar{L}_{\alpha\beta} - L_{\alpha\beta} \qquad \textit{according to (2.3–18b)} \qquad (2.3\text{–}30b)$$

the first and second partial derivations of

$$z_{i.} = x_{i,a}\, z^a = c_{i.a}\, z^a \qquad (2.3\text{–}31a)$$

are needed in addition to (2.3–29) with respect to the surface coordinates (u^α) as functions of the derivations of the q^a. These result in

$$\begin{aligned} z_{i,\alpha} &= c_{i.c}\,(z^c_{,\alpha} + \Gamma^c_{ab}\, q^b_{,\alpha}\, z^a) \\ z_{i,\alpha\beta} &= c_{i.e}\,\Big\{ z^e_{,\alpha\beta} + \Gamma^e_{ab}\,(q^b_{,\beta}\, z^a_{,\alpha} + q^b_{,\alpha}\, z^a_{,\beta}) \\ &\quad + \Big[\Gamma^e_{ab}\, q^b_{,\alpha\beta} + (\Gamma^e_{ab,d} + \Gamma^e_{cd}\,\Gamma^c_{ab})\, q^b_{,\alpha}\, q^d_{,\beta}\Big]\, z^a \Big\} \,. \end{aligned} \qquad (2.3\text{–}31b)$$

Owing to (2.3–10d) for the second partial derivations of the q^c with respect to the surface coordinates one obtains

$$q^1_{,\alpha\beta} = q^2_{,\alpha\beta} = 0 \qquad (2.3\text{–}31c)$$

$$q^3_{,\alpha\beta} = H_{,11}\delta^1_\alpha\delta^1_\beta + H_{,12}\delta^1_\alpha\delta^2_\beta + H_{,21}\delta^2_\alpha\delta^1_\beta + H_{,22}\delta^2_\alpha\delta^2_\beta \,. \qquad (2.3\text{–}31d)$$

Taking (2.3–29–31) into account, all curvature computations (2.3–19–21) can be carried out in the external surface normal coordinates q^a.

To compute the **rotations of the surface elements** (2.3–22–28) one starts again from the assumptions (2.3–8, 9a, b). Contrary to the reflections on the *section 2.3.2.3* the h–component in the direction of the surface normals must here also be considered with the internal surface normal coordinate system. The use of the external q^a–system as well as of the internal p^a–system necessitates providing the indices of the latter generally with a tilde ~ . Thus, the following representations apply:

$$b_{i.a} = \partial x_{i.}/\partial p^a = c_{i.c}\, q^c_{,\tilde{a}} \;, \quad b^{\tilde{a}}_{i.} = \partial p^a/\partial x_{i.} = c^c_{i.}\, p^a_{,c} \;, \tag{2.3–32a}$$

and owing to (2.3–31a) one obtains

$$z_{i,\tilde{b}} = c_{i.d}\,(\, z^d_{,\tilde{b}} + \Gamma^d_{ef}\, q^e_{,\tilde{b}}\, z^f\,) =: c_{i.d}\,\delta^\beta_b\,(\, z^d_{,\tilde{\beta}} + \Gamma^d_{ef}\, q^e_{,\tilde{\beta}}\, z^f\,)\,. \tag{2.3–32b}$$

The second form meets $z_{i,\tilde{3}} = 0$ according to (2.3–22b). For this it follows

$$z^d_{,\tilde{3}} = -\Gamma^d_{ef}\, q^e_{,\tilde{3}}\, z^f = -\Gamma^d_{ef}\, n^e\, z^f\;. \tag{2.3–32c}$$

To (2.3–9c) corresponds here the equation

$$b_{i.a}\, z_{i,\tilde{b}} = q^c_{,\tilde{a}}\,\delta^\beta_b\,(\, z^d_{,\tilde{\beta}} + \Gamma^d_{ef}\, q^e_{,\tilde{\beta}}\, z^f\,)\, g_{cd}\;. \tag{2.3–32d}$$

If it is put in (2.3–24a, 25a) one obtains for the

symmetric portions of the displacement tensor

$$\begin{aligned}\tilde{\varepsilon}_{ab} \;=\;& [\, q^c_{,\tilde{a}}\,\delta^\beta_b\,(\, z^d_{,\tilde{\beta}} + \Gamma^d_{ef}\, q^e_{,\tilde{\beta}}\, z^f\,) \\ &+ q^c_{,\tilde{b}}\,\delta^\alpha_a\,(\, z^d_{,\tilde{\alpha}} + \Gamma^d_{ef}\, q^e_{,\tilde{\alpha}}\, z^f\,)\,]\, g_{cd}/2\end{aligned} \tag{2.3–33a}$$

and for the

tensor of rotation

$$\begin{aligned}\tilde{\xi}_{ab} \;=\;& [\, q^c_{,\tilde{a}}\,\delta^\beta_b\,(\, z^d_{,\tilde{\beta}} + \Gamma^d_{ef}\, q^e_{,\tilde{\beta}}\, z^f\,) \\ &- q^c_{,\tilde{b}}\,\delta^\alpha_a\,(\, z^d_{,\tilde{\alpha}} + \Gamma^d_{ef}\, q^e_{,\tilde{\alpha}}\, z^f\,)\,]\, g_{cd}/2\,.\end{aligned} \tag{2.3–33b}$$

One starts from the assumption (2.1–8d), so

$$q^\alpha \equiv u^\alpha \equiv p^\alpha\;, \tag{2.3–34a}$$

with which the transformation matrices between the q^a–system and the p^a–system result as follows:

$$q^d_{,\tilde{a}} = c^d_{i.}\, b_{i.\tilde{a}}\;, \tag{2.3–34b}$$

especially:

$$q^{\gamma}_{,\bar{\alpha}} = \delta^{\gamma}_{\alpha} \; , \quad q^{3}_{,\bar{\alpha}} = H_{,\alpha} \; ,$$
$$q^{d}_{,\bar{3}} \;=\; c^{d}_{i.} n_{i.} \;=\; n^{d} \;=\; g^{cd} \varepsilon_{abc} q^{a}_{,\bar{1}} q^{b}_{,\bar{2}} f^{-1/2} \qquad (2.3\text{–}34c)$$

with

$$f_{\alpha\beta} = q^{a}_{,\bar{\alpha}} q^{b}_{,\bar{\beta}} g_{ab} \; , \qquad f = |f_{\alpha\beta}| = f_{11} f_{22} - f_{12}^{2} \; , \qquad (2.3\text{–}34d)$$

and

$n_{i.} = b_{i.\bar{3}} =$ *normal of the reference surface of the* p^a*–system* .

For the transformation matrices between the p^a–system and the q^a–system inverse to this applies

$$p^{d}_{,a} = b^{\bar{d}}_{i.} c_{i.a} \; , \qquad p^{d}_{,a} q^{a}_{,\bar{b}} = \delta^{d}_{b} \; , \qquad (2.3\text{–}35a)$$

especially

$$p^{\gamma}_{,\alpha} = \delta^{\gamma}_{\alpha} \; , \quad p^{3}_{,\alpha} = h_{,\alpha} = -H_{,\alpha}/n^{3} \; ,$$
$$p^{d}_{,3} = b^{\bar{d}}_{i.} m_{i.} = \tilde{m}^{d} = (\, 0 \; , \; 0 \; , \; 1/n^{3}) \qquad (2.3\text{–}35b)$$

with

$m_{i.} = c_{i.3} =$ *normal of the reference surface of the* q^a*–system* ,
$n_{i.}$ *according to (2.3–34d)* .

The real difference as to the results (2.3–10b, c) lies in the fact that the transformation matrices $q^{d}_{,\bar{3}}$ are taken into account. If one considers only the 'linear surface deformation tensor' (2.3–24b), which will generally be the case, (2.3–33a) goes over into (2.3–10c).

However, for the 'rotation tensor' (2.3–33b) the three-dimensional representation in the p^a–system is important because only this representation allows the complete representation of the

infinitesimal rotation vector (2.3–27b):

$$\begin{aligned} \tilde{d}^{c} &= (1/2)\, \tilde{\varepsilon}^{abc} \tilde{\xi}_{ab} = (1/2)\, \tilde{\varepsilon}^{\alpha bc} b_{i.\bar{b}} z_{i,\bar{\alpha}} \\ &= (1/2)\, \tilde{\varepsilon}^{\alpha bc} q^{g}_{,\bar{b}} (\, z^{d}_{,\bar{\alpha}} + \Gamma^{d}_{ef} q^{e}_{,\bar{\alpha}} z^{f} \,) g_{dg} \; . \end{aligned} \qquad (2.3\text{–}36a)$$

In the case of restriction to the 'normal rotation tensor' (2.3–25b), only the normal component of the rotation vector (2.3–36a)

$$\tilde{d}^{3} = (1/2)\, \tilde{\varepsilon}^{\alpha\beta 3} q^{g}_{,\bar{\beta}} (\, z^{d}_{,\bar{\alpha}} + \Gamma^{d}_{ef} q^{e}_{,\bar{\alpha}} z^{f} \,) g_{dg} \qquad (2.3\text{–}36b)$$

has to be computed. The **covariant derivations of the contravariant rotation vector** (2.3–36a) corresponding to (2.3–28b) can generally be computed in the form

$$\tilde{d}^{c}_{;\tilde{\beta}} = (1/2)\,\tilde{\varepsilon}^{abc}\left[q^{g}_{,\tilde{b}}\,(\,z^{d}_{,\tilde{a}} + \Gamma^{d}_{ef}\,q^{e}_{,\tilde{a}}\,z^{f}\,)\,g_{dg}\right]_{;\tilde{\beta}} \ .$$

The result must be identical with (2.3–28b) if there the Cartesian coordinates are eliminated. In this way first

$$\tilde{d}^{c}_{;\tilde{\beta}} = (1/2)\,\tilde{\varepsilon}^{mnc}[(-\tilde{\Gamma}^{\alpha}_{\beta m} + \tilde{L}^{\alpha}_{\beta}\,\delta^{3}_{m})\,b_{j.n}\,z_{j,\tilde{\alpha}} + \delta^{\alpha}_{m}\,b_{j.n}\,z_{j,\tilde{\alpha}\tilde{\beta}}] \qquad (2.3\text{–}37a)$$

is obtained, in which according to (2.3–32d) is

$$b_{i.n}\,z_{i,\tilde{\alpha}} = q^{c}_{,\tilde{n}}\,(\,z^{d}_{,\tilde{\alpha}} + \Gamma^{d}_{ef}\,q^{e}_{,\tilde{\alpha}}\,z^{f}\,)\,g_{cd} \ , \qquad (2.3\text{–}37b)$$

and starting from (2.3–32b) one obtains

$$\begin{aligned} b_{i.n}\,z_{i,\tilde{\alpha}\tilde{\beta}} &= q^{c}_{,\tilde{n}}\,(\,z^{d}_{,\tilde{\alpha}\tilde{\beta}} + \Gamma^{d}_{ef}\,q^{e}_{,\tilde{\alpha}}\,z^{f}_{,\tilde{\beta}} + \Gamma^{d}_{ef}\,q^{e}_{,\tilde{\alpha}\tilde{\beta}}\,z^{f} + \Gamma^{d}_{ef,o}\,q^{e}_{,\tilde{\alpha}}\,q^{o}_{,\tilde{\beta}}z^{f}\,)\,g_{cd} \\ &+ \Gamma^{c}_{do}\,q^{o}_{,\tilde{\beta}}\,p^{g}_{,c}\,(\,z^{d}_{,\tilde{\alpha}} + \Gamma^{d}_{ef}\,q^{e}_{,\tilde{\alpha}}\,z^{f}\,)\,\tilde{f}_{ng} \ . \qquad (2.3\text{–}37c) \end{aligned}$$

(2.3–37b, c) are to be substituted in (2.3–37a) and the transformation matrices are to be computed according to (2.3–34, 35).

2.3.3.3 Computations in Ellipsoidal Coordinates

For the representation of the external surface deformation theory in geographical ellipsoidal coordinates, it is started from the results of *section 2.3.2.4*. The partial derivations of the ellipsoidal coordinates $q^a = (\lambda, \phi, H)$ according to the internal surface normal coordinates $\tilde{p}^a = (\lambda, \phi, h)$ result in conformity with (2.3–34b, c). By this and taking into account the Christoffel symbols of the 2nd kind (2.2–13) all quantities (2.3–30) pp., (2.3–36) pp. relevant to deformation studies can be computed. A further solution of the tensor multiplication offers no advantages for programming because the Christoffel symbols have only relatively few zero components.

2.3.4 Deformations and Stresses

In this paper the general aspects of the solution of boundary value problems of elastostatics are not examined. In the following only a short insight into the **local stress–deformation relations at the Earth's surface** is given, which show quite interesting correlations between the internal and external deformation measures of the surface F and pertinent stresses and changes in stresses.

All reflections refer to a test point $P \in F$, and they are realized in the **internal surface coordinate system** $\tilde{\mathrm{S}}$ spanned in P according to (2.1–8)

$$p^a = (u^\alpha, h)^a \,, \qquad u^\alpha = \textit{surface coordinates of the reference surface} \tag{2.3–38a}$$

with the metric tensors

$$\tilde{f}_{ab} = x_{i,\tilde{a}}\, x_{i,\tilde{b}} \,, \qquad \tilde{f}_{\alpha\beta} = f_{\alpha\beta} = x_{i,\alpha}\, x_{i,\beta} \,. \tag{2.3–38b}$$

The computations simplify themselves considerably if in this local system $\tilde{S}$ **Cartesian coordinates**

$$\begin{aligned} &\tilde{x}_{i.} =: x_{i.} \,, \\ &\textit{axis 1. in the direction of the } u^1 - \textit{line} \,, \\ &\textit{axis 2. in the tangential plane of } F \,, \\ &\textit{axis 3. in the direction of the surface normals } \tilde{n}_{i.} = \delta_{i3.} \end{aligned} \tag{2.3–39a}$$

are introduced. In the following the tilde ~ over the vector symbols is generally omitted in as far as this may lead to confusion. The transformation matrices are

$$\begin{aligned} x_{i,1} &= f_{11}^{1/2}\, \delta_{i1.} \,, \\ x_{i,2} &= (\, f_{12} \,,\, f^{1/2} \,,\, 0\,)_{i.}\, f_{11}^{-1/2} \,, \\ x_{i,3} &= \delta_{i3.} \\ p^a_{,1} &= f_{11}^{-1/2}\, \delta^a_1 \,, \\ p^a_{,2} &= (\, f_{12} \,,\, -f_{11} \,,\, 0\,)^a\, (f\, f_{11})^{-1/2} \,, \\ p^a_{,3} &= \delta^a_3 \,, \end{aligned} \tag{2.3–39b}$$

with

$$f = f_{11}\, f_{22} - f_{12}^2 \,. \tag{2.3–39c}$$

If the **internal surface deformations** have been determined, the surface–deformation tensor $\tilde{\varepsilon}_{\gamma\delta}$, (2.3–6b) is known and can be transformed by means of (2.3–39b) into the Cartesian local system:

$$\varepsilon_{\alpha\beta.} = p^{\gamma}_{\alpha}\, p^{\delta}_{\beta}\, \tilde{\varepsilon}_{\gamma\delta} \ , \qquad \alpha,\, \beta,\, \gamma,\, \delta \in \{1,\, 2\} \ . \tag{2.3–40a}$$

With this the three Cartesian components $\varepsilon_{11.}$, $\varepsilon_{12.}$, $\varepsilon_{22.}$ can be assumed as given. A further precondition consists of the disappearance of the surface stresses

$$s_{i.3} = \sigma_{ij.}\, n_{j.} = \sigma_{i3.} = 0 \ ; \tag{2.3–40b}$$

Concerning the dynamic fundamentals needed here and in the following, please (cf. Heitz 1980–1983).

An isotropic, linear–elastic state is assumed, for which the parameters of elasticity

$$\begin{aligned} &\lambda,\, \mu = \textit{Lamé's parameters} \ , \\ &\mu = \textit{shear modulus} \ , \\ &\nu = \textit{Poisson's parameter}, \\ &E = 2\,\mu\,(1+\nu) = \textit{Young's parameter}, \end{aligned} \tag{2.3–41a}$$

are used. Thus, **Hooke's law** according to Heitz (1980–1983) (8–33, 49) reads

$$\begin{aligned} \sigma_{ij.} &= 2\,\mu\,\varepsilon_{ij.} + \lambda\,\varepsilon_{kk.}\,\delta_{ij.} \\ &= 2\,\mu\,\{\,\varepsilon_{ij.} + [\nu/(1-2\,\nu)]\,\varepsilon_{kk.}\,\delta_{ij.}\,\} \ . \end{aligned} \tag{2.3–41b}$$

From this follows the combination with (2.3–40b)

$$\sigma_{i3.} = 0 = 2\,\mu\,\varepsilon_{i3.} + \lambda\,\varepsilon_{kk.}\,\delta_{i3.} \ . \tag{2.3–41c}$$

A direct result is

$$\varepsilon_{13.} = \varepsilon_{23.} = 0 \ , \tag{2.3–41d}$$

so that the surface normal constitutes a **principal deformation and stress direction.** The two other principal directions thus fall on the tangential plane 1.–2., and in addition to (2.3–41d)

$$\varepsilon_{12.} = 0 \tag{2.3–41e}$$

is assumed, which is always possible by an appropriate orientation of the coordinate axes 1., 2., then the principal axes coincide with these axes. Further, one obtains from (2.3–41c)

$$\varepsilon_{33.} = [\lambda/(2\,\mu)]\,\bar{q}_V = [\nu/(1-\nu)]\,\bar{q} \qquad (2.3\text{–}41f)$$

with the volume dilatation $\bar{q}_V$ and the surface dilatation $\bar{q}$:

$$\bar{q}_V = \varepsilon_{kk.} = [2\,\mu/(\lambda + 2\,\mu)]\,\bar{q}\ ,\bar{q} = \varepsilon_{\alpha\alpha.} = \varepsilon_{11.} + \varepsilon_{22.}\ . \qquad (2.3\text{–}41g)$$

With these results the **horizontal stresses** can be computed on the basis of Hooke's law, and it is

$$\begin{aligned} s_{i.1} &= \sigma_{i1.} = 2\,\mu\,\{\,\varepsilon_{11.} + [\nu/(1-\nu)]\,\bar{q}\ ,\ \ \varepsilon_{12.}\ ,\ \ 0\,\}_{i.} \\ s_{i.2} &= \sigma_{i2.} = 2\,\mu\,\{\,\varepsilon_{12.}\ ,\ \ \varepsilon_{22.} + [\nu/(1-\nu)]\,\bar{q}\ ,\ \ 0\,\}_{i.}\ . \end{aligned} \qquad (2.3\text{–}41h)$$

On the condition of (2.3–41e), it applies that

$$s_{1.2} = s_{2.1} = 0\ . \qquad (2.3\text{–}41i)$$

Now follow computations of **changes of stresses at the Earth's surface F**, which result from the deformation $F \Rightarrow \bar{F}$. To do this one starts from the differential equations of elastostatics in the simplest form in the Cartesian local system $\tilde{S}$ (2.3–39a)

$$\sigma_{ij,j} = 0_{i.} \qquad (2.3\text{–}42a)$$

This corresponds to Heitz (1980–1983) (8–105d) neglecting gravity changes due to the deformations. Dissolved according to normal derivations (2.3–42a) reads

$$\partial s_{i.3}/\partial x_{3.} = \sigma_{i3,3} = -\sigma_{i1,1} - \sigma_{i2,2}\ . \qquad (2.3\text{–}42b)$$

The derivations of the right side are to be formed on the basis of Hooke's law (2.3–41b):

$$\sigma_{i1,\alpha} = 2\,\mu\,\{\,\varepsilon_{i1,\alpha} + [\nu/(1-\nu)]\,\bar{q}_{,\alpha}\,\delta_{i\alpha.}\,\}\ ,\quad \alpha \in \{1,\,2\} \qquad (2.3\text{–}43a)$$

with

$$\bar{q}_{,\alpha} = \varepsilon_{11,\alpha} + \varepsilon_{22,\alpha}\ . \qquad (2.3\text{–}43b)$$

When substituted into (2.3–42b) it results that

$$\sigma_{i3,3} = -2\,\mu\,\{\varepsilon_{i1,1} + \varepsilon_{i2,2} + [\nu/(1-\nu)]\,(\,\bar{q}_{,1}\,\delta_{i1.} + \bar{q}_{,2}\,\delta_{i2.}\,)\}\ . \qquad (2.3\text{–}43c)$$

So far all formulas are rigorous. For the further evaluation of $\sigma_{i3,3}$ the **approximations are done in the neighbourhood of the test point P**

$$F \equiv \textit{tangential plane of } F \textit{ in } P\,, \qquad \bar{F} = \textit{curved surface}, \tag{2.3–44a}$$

which are justified with the assumption of small displacements (2.3–6a). Hence, the following simplifications apply

$$f_{\alpha\beta} = \delta_{\alpha\beta.}\,, \qquad f = 1\,, \qquad x_{i,\alpha} = \delta_{i,\alpha.}\,, \qquad x_{i,\alpha\beta} = 0\,, \qquad L_{\alpha\beta} = 0\,; \tag{2.3–44b}$$

$$\varepsilon_{i\alpha.} = (z_{i,\alpha} + z_{\alpha,i})/2\,, \qquad \varepsilon_{i\alpha,\beta} = (z_{i,\alpha\beta} + z_{\alpha,i\beta})/2\,, \qquad z_{i,3} = 0\,, \qquad z_{i,3\alpha} = 0\,; \tag{2.3–44c}$$

$$\bar{f}_{\alpha\beta} = \delta_{\alpha\beta.} + z_{\alpha,\beta} + z_{\beta,\alpha} = \delta_{\alpha\beta.} + 2\,\varepsilon_{\alpha\beta.}\,, \qquad \delta f = 2\,(z_{1,1} + z_{2,2}) = 2\,(\varepsilon_{11.} + \varepsilon_{22.}) \qquad \bar{L}_{\alpha\beta} = \delta L_{\alpha\beta} = z_{3,\alpha\beta}\,. \tag{2.3–44d}$$

If the axes 1., 2. are identified with the principal directions of curvature, then it applies that

$$\delta L_{12} = z_{3,12} = 0\,, \tag{2.3–45a}$$

and owing to (2.3–21c–e) one obtains

$$\delta\kappa_{N(n)} = \delta L_{(nn)} = z_{3,(nn)}\,, \tag{2.3–45b}$$

$$\delta H = (\delta L_{11} + \delta L_{22})/2 = (z_{3,11} + z_{3,22})/2\,. \tag{2.3–45c}$$

With this the components of (2.3–43c) can be represented as follows:

$$\sigma_{13,3} = -2\,\mu\,\{\,\varepsilon_{11,1} + \varepsilon_{12,2} + [\nu/(1-\nu)]\,\bar{q}_{,1}\,\}\,, \qquad \sigma_{23,3} = -2\,\mu\,\{\,\varepsilon_{12,1} + \varepsilon_{22,2} + [\nu/(1-\nu)]\,\bar{q}_{,2}\,\}\,, \tag{2.3–46a}$$

$$\sigma_{33,3} = -2\,\mu\,\delta H\,, \qquad 2\,\mu = E/(1+\nu)\,. \tag{2.3–46b}$$

The **horizontal components** $\sigma_{i3,3}$, $i \in \{1, 2\}$ are determined only by the internal surface deformations. The gradients needed in (2.3–46a) of the Cartesian surface–deformation tensor are obtained on the basis of the 'curvilinear components' $\tilde{\varepsilon}_{\gamma\delta}$, (2.3–10c) determined originally, as a rule, through transformation according to (2.3–40a) in the following way:

$$\begin{aligned} \varepsilon_{\alpha\beta,\mu} &= p^{\gamma}_{\alpha}\, p^{\delta}_{\beta}\, p^{\nu}_{\mu}\, \tilde{\varepsilon}_{\gamma\delta;\nu} \;, \\ &= p^{\gamma}_{\alpha}\, p^{\delta}_{\beta}\, p^{\nu}_{\mu}\, [\, \tilde{\varepsilon}_{\gamma\delta,\nu} - \tilde{\Gamma}^{\epsilon}_{\gamma\nu}\, \tilde{\varepsilon}_{\epsilon\delta} - \tilde{\Gamma}^{\epsilon}_{\delta\nu}\, \tilde{\varepsilon}_{\epsilon\gamma}\,] \;, \end{aligned} \tag{2.3–47a}$$

with $\alpha,\ \beta,\ \ldots\ \epsilon \in \{1,\ 2\}$.

The **vertical component** $\sigma_{33,3}$, (2.3–46b) is a function of the change of the mean curvature (2.3–21d)

$$\delta H = \bar{H} - H \;, \tag{2.3–47b}$$

so that it is determined by the external surface deformations.

Chapter 3

Geometric Modelling

3.1 General Fundamentals

In *chapter 1: Introduction* the importance regarding (1–5) and (1–7) of geometric modelling of surfaces and their temporal variations by interpolation functions for the observation quantities in discretely distributed observation stations (1–6), namely the **point and displacement coordinates**

$$q^a = q^a(u^\alpha) \ , \qquad\qquad z^a = z^a(u^\alpha) \qquad\qquad (3.1\text{–}1a)$$

has already been pointed out. Here, only surface normal coordinates (2.1–6) are used, which includes also the special Gaussian surface representation in Cartesian coordinates (2.1–4). With this and taking (2.3–8) into account, (3.1–1a) are to be substituted by

$$H = H(u^\alpha) \ , \qquad\qquad z^a = z^a(u^\alpha) \ . \qquad\qquad (3.1\text{–}1b)$$

These **height and displacement coordinates** are in the following generally referred to as

$$y(u^\alpha) \in \{H(u^\alpha) \ , \ z^a(u^\alpha)\} \qquad\qquad (3.1\text{–}1c)$$

and the special designations

$$\begin{aligned} P_q \ &= \ \textit{points of source} \qquad\qquad (3.1\text{–}2a)\\ &= \ \textit{stations with observed values } y_q \ , \end{aligned}$$

$$\begin{aligned} P_p \ &= \ \textit{test points} \qquad\qquad (3.1\text{–}2b)\\ &= \ \textit{stations with values to be interpolated } \hat{y}_p := \hat{y}(u_p^\alpha) \end{aligned}$$

are introduced. The determination of an interpolation function $\hat{y}(u^\alpha)$, (3.1–2b), with the points of support (3.1–2a) is described as **interpolation or approximation**, depending on whether the y_q–values are exactly observed or approximated only to minimize residuals. Here, the designation 'interpolation' is generally used for both cases. If besides the observations (3.1–2a) there are no additional data for the field quantities $y(u^\alpha)$ to be interpolated, e.g. in the form of dynamic models, which is assumed here in any case, purely **geometrical modellings or interpolation procedures** are concerned. These must in general meet the following conditions:

- *the* ***observation stations*** (3.1–3a)
  or points of source (3.1–2a) are to be selected so densely that a linear interpolation between neighbouring stations ensures the accuracy achievable or aimed at
- *the selected* ***geometric interpolation function*** (3.1–3b)
  must according to (1–7b) ensure an approximately linear interpolation between neighbouring observation stations.

The **accuracy achievable** with a given density of observation or source points can be estimated relatively simply. For individual test points linear interpolations on the basis of differently selected, neighbouring 'source triangles' are carried out. The differences of the interpolation results following from this one and the same test point are a measure for the accuracy of the approximation. If it is not sufficient for the particular formulation of the problem of geometric modelling, an appropriate densification of the observation points must be made.

Basically, it is possible to meet very well the requirements (3.1–3a, b) through

- **Graphic interpolation,** (3.1–4a)
  on the basis of the isolines drawn in a map of the function values $y(u^\alpha)$.

However, such *'linearly discretized interpolation functions'* are not computer–compatible. Therefore, they are substituted by completely or piece by piece

- **Analytic interpolation functions,** (3.1–4b)
  which besides having to fulfil (3.1–3) must be two times continuous according to the u^α in order to allow the completely analytic computation of the internal and external surface deformation measures,

i.e. according to (1–7b).

Here, only **linear interpolation procedures** as to (3.1–4b) are considered. With (3.1–2) the result of a linear interpolation can generally be represented in the form

$$\hat{y}_p = y_p^T + \sum_q a_{pq}\,(y_q - y_q^T) \tag{3.1–5a}$$

where

$$\begin{aligned} a_{pq} &= \textit{coefficients} \\ &= \textit{functions to be determined of } u_p^\alpha,\ \{y_q,\ u_q^\alpha\}\ , \end{aligned} \tag{3.1–5b}$$

$$\begin{aligned} y_p^T &= \textit{trend function} \\ &= \textit{known function of } u_p^\alpha\ . \end{aligned} \tag{3.1–5c}$$

As **trend functions** polynomials of relatively low degree n′ enter above all into consideration:

$$y_p^T = y_{p0}^T + \sum_{n_1=1}^{n_1'} \sum_{n_2=1}^{n_2'} b_{n_1 n_2}(u^1)^{n_1}(u^2)^{n_2}\ , \quad \textit{with order } n' = n_1' + n_2'\ . \tag{3.1–6a}$$

For smaller interpolation areas the mean value

$$y_p^T = y_{p0}^T = (\sum_q y_q)/n_q\ , \quad n_q = \textit{number of the } P_q \tag{3.1–6b}$$

is often a suitable trend function.

Various linear interpolation or approximation methods differ from each other only by the determination of the coefficients a_{pq} in (3.1–5a). If for different methods the conditions (3.1–3) are complied with in the same quality, the results for the coefficients (3.1–5b) can in each case differ only slightly for one and the same test point P_p.

A good survey of the analytic interpolation methods (3.1–4b) can be found in Abramowsk and Müller (1991). In any case, **polynomial interpolation** must be mentioned first, which is described in *chapter 3.2.* This is formally the simplest method. The oscillations with higher degrees of polynomials has a disadvantageous effect which may lead to coarse violations of the approximation requirement (3.1–3b). This may be counteracted by, among others, the **interpolation by spline functions.** Spline functions are piece by piece continuous polynomials of a low degree (e.g. bicubic) over rectangular sectors of the coordinates, for which at the boundaries continuity in the first, second or higher derivations according to the coordinates is required (cf. Abramowsk and Müller 1991, Chapter 4 and 5). The spline technique is formally relatively complex and depends on a regular source point grid. A procedure more advantageous for the problems under consideration here constitutes the **interpolation by collocation**, dealt with in *section 3.3*, called also interpolation according to the least squares. The covariances of the function values assumed as being known, which are generally considered as solely distance–dependant, form the basis of this method.

The spline and collocation techniques constitute always interpolation methods in the stricter sense of the remarks after (3.1–2b): in the source points applies: $\hat{y}_q = y_q$. On the contrary, polynomial interpolation is often used as a method of approximation.

The **subdivision by finite triangles** treated in *section 3.4* constitutes a special case of the interpolation methods, with which the surface F is decomposed into triangles with the corner points coinciding with the observation stations (3.1–2b). For each triangle separate interpolation functions (3.1–5) are used.

3.2 Interpolation by Polynomials

A point field

$$\{P\} \in \textit{region } F \textit{ of the Earth's surface } O \qquad (3.2\text{–}1a)$$

was observed at two points of time

t *and*

As a result **surface normal coordinates** (2.3–8) are available, i.e.

$$q^a(u^\alpha) = [u^1, u^2, H(u^\alpha)]^a ,$$

$$\bar{q}^a(u^\alpha) = [\bar{u}^1(u^\alpha), \bar{u}^2(u^\alpha), \bar{H}(u^\alpha)]^a , \qquad (3.2\text{–}2a)$$

and thus the **displacement coordinates**

$$z^a(u^\alpha) =: z^a(u^\alpha) = \bar{q}^a(u^\alpha) - q^a(u^\alpha) . \qquad (3.2\text{–}2b)$$

For the region F of the Earth's surface and the displacement field observed, one starts from analytical functions in the form of a

Polynomial for the height $q^3 \equiv H$:

$$H(u^\alpha) \approx \sum_{n_1=0}^{n_1'} \sum_{n_2=0}^{n_2'} h_{n_1 n_2} (u^1)^{n_1} (u^2)^{n_2} \qquad (3.2\text{–}3a)$$

and of

Polynomials for the displacement coordinates z^a :

$$z^a(u^\alpha) \approx \sum_{n_1=0}^{n_1'} \sum_{n_2=0}^{n_2'} c^a_{n_1 n_2} (u^1)^{n_1} (u^2)^{n_2} . \qquad (3.2\text{–}3b)$$

The **Determination of Coefficients**

$$h_{n_1 n_2} \quad and \quad c^a_{n_1 n_2} \qquad (3.2\text{–}3c)$$

is done on the basis of the $H-$ and z^a-values (3.2–2a, b), i.e. generally by a formal application of the adjustment according to the least squares method. After the determination of the coefficients (3.2–3c) all measuring quantities treated in *sections 2.3.2 and 2.3.3* can be computed with regard to the internal and external surface deformations.

The **degree of the polynomials**

$$n' = n_1' + n_2' \qquad (3.2\text{–}4a)$$

is composed of the degrees of development differently assumable in the two surface coordinates

$$n_1' = \textit{degree in } u^1 , \qquad n_2' = \textit{degree in } u^2 . \qquad (3.2\text{–}4b)$$

If n'_{gr} is the larger and n'_{kl} the smaller one of the degree numbers n'_1, n_2 the number of coefficients amounts to:

$$N_{n_1 n_2} = (n'_{kl} + 1)\,[\,(n'_{kl} + 2)/2 + n'_{gr} - n'_{kl}\,]\ , \qquad n'_{gr} \geq n'_{kl}\ . \quad (3.2\text{–}4c)$$

A special problem of the representation of polynomials consists in that they tend to **oscillations between the source points**, which contradicts the requirement of linearity (3.1–3b). This occurs particularly with higher degrees of polynomials, and no generally valid rules can be given to avoid oscillations. The following requirements are to be observed in any case:

- *the* ***observation stations*** (3.2–5a)
  or source points (3.1–2a) are to be selected nearly equidistantly and
- *the* ***degree of development of the polynomials n ′*** (3.2–5b)
  is to be selected so low that an over–determination as large as possible of the coefficients (3.2–3c) is ensured.

Both postulates present no difficulties as regards the interpolation of the terrain model represented by the H–values since this is subject only to relatively low accuracy requirements. The H–values can in principle be taken from topographic maps at problem–adapted scales. With the z^a–values which, as a rule, have to be determined very precisely, a 'densification of the observation' can be obtained without further observations just by adding

- **linearly interpolated $\bar{z}^a$–values** (3.2–5c)
  for source points $\bar{P}_q$, between neighbouring observation stations P_q (triangular or rectangular meshes).

The polynomial degrees are to be selected such that the number of coefficients (3.2–4b) becomes

$$N_{n_1 n_2} \leq \textit{number of observation stations } P_q\ . \quad (3.2\text{–}5d)$$

3.3 Interpolation by Collocation

The interpolation by collocation method was applied for the first time by *Moritz (1973)* to geodetic problems as 'Least Squares Prediction' and has since developed into a mathematical instrument widely used in geodesy (cf. especially Moritz 1973, 1980; Moritz and Sünkel 1977). Here, the collocation is needed only in its original form as 'interpolation according to least squares', the essential features of which are given in the following.

This theory starts from the interpolation formulation (3.1–5), which can also be interpreted as linear **autocorrelation**. With the assumptions

$$\hat{y}_p := \hat{y}_p - \hat{y}_p^T \,, \qquad y_q := y_q - y_q^T \tag{3.3–1a}$$

the linear autocorrelation of the $\hat{y}_p$, y_q reads generally

$$\hat{y}_p = \sum_{q=1}^{n_q} a_{pq}\, y_q \,, \qquad n_q = \textit{number of source points} \,. \tag{3.3–1b}$$

If a trend function in the form of a polynomial (3.1–6) is introduced where coefficients are determined according to the method of least squares, then it applies with the designation generally used in the following

$$< f_k > := (\sum_{k=1}^{n} f_k)/n \quad = \quad \textit{mean of the } f_k \ : \tag{3.3–1c}$$

$$< y_q > = 0 \qquad \Rightarrow < \hat{y}_p > \approx 0 \,. \tag{3.3–1d}$$

For the functions (3.1–1b) to be interpolated, the autocorrelation formulations (3.3–1b) have the special forms

$$\hat{H}_p = \sum_{q=1}^{n_q} a_{pq}\, H_q \,, \qquad H := H - H^T \,, \tag{3.3–2a}$$

$$(\hat{z}^a)_p = \sum_{q=1}^{n_q} (b^a_c)_{pq}\, (z^c)_q \,, \qquad z^a := z^a - (z^a)^T \,. \tag{3.3–2b}$$

(3.3–2b) represents a linear vector function which brings about a correlation of all three components $(z^a)_p$, respectively with all three components $(z^c)_q$ in the source points, which is also designated as 'cross correlation'. However, this complex approach can, as a rule, be reduced to a simple autocorrelation of the single vector components among themselves:

$$(\hat{z}^a)_p = \sum_{q=1}^{n_q} b_{pq}\,(z^a)_q\ . \tag{3.3–2c}$$

The law of autocorrelation is in any case to be selected such that the coefficients $(b^a_c)_{pq}$ tend asymptotically to zero with increasing distance $P_p - P_q$ [cf. (3.3–5)] and that the three source points P_p located nearest to the test point P_q have the main influence on $(\hat{z}^a)_p$. If one assumes for this 'source triangle' a (nearly) linear interpolation according to the postulate (3.1–3b), then the correlation formulation (3.3–2c) [cf. *section 3.4.2, (3.4–4b)*] results for this, which is certainly also justified with respect to the source points lying outside the neighbouring source triangle.

For the **determination of the correlation coefficient** of (3.3–2a, c) through collocation or interpolation according to least squares, one may start from the general formulation (3.3–1a, b). If y_p are the true function values in the test points P_p, then it results for the approximation errors

$$\delta y_p := y_p - \hat{y}_p = y_p - \sum_{q=1}^{n_q} a_{pq}\, y_q\ . \tag{3.3–3a}$$

Their mean square sum can be represented by introducing the **covariance functions**

$$C_{pq} = C_{qp} := \langle y_p\, y_q \rangle\ ,\quad C_0 := \langle y_p\, y_p \rangle \approx \langle y_q\, y_q \rangle \tag{3.3–3b}$$

in the form

$$\begin{aligned} (M_p)^2 &= \langle \delta y_p\, \delta y_p \rangle \\ &= C_0 - 2\sum_{q=1}^{n_q} a_{pq}\, C_{pq} + \sum_{q=1}^{n_q}\sum_{l=1}^{n_q} a_{pq}\, a_{pl}\, C_{ql}\ , \end{aligned} \tag{3.3–3c}$$

$q,\ l$ = *source point indices* , n_q = *number of source points* .

The **Least–Squares–Postulate**

$$(M_p)^2 \rightarrow \textit{minimum} \tag{3.3–4a}$$

leads to the conditional equations

$$\partial (M_p)^2 / \partial a_{pq} = 0 \tag{3.3–4b}$$

for the coefficients a_{pq}, which when applied to (3.3–3c) read fully:

$$\sum_{q=1}^{n_q} a_{pq}\, C_{ql} = C_{pl} \ , \qquad l = 1,\ 2, \ldots n_q \ . \tag{3.3–4c}$$

This is a system of n_q linear equations with a symmetrical matrix of coefficients $C_{pq} = C_{qp}$, by which the n_q coefficients a_{pq} are in general determined unambiguously:

$$\begin{aligned} a_{pq} &= \sum_{l=1}^{n_q} C_{ql}^{-1}\, C_{pl} \ , \\ C_{ql}^{-1} &= \textit{inverse of the covariance matrix } C_{ql} \, . \end{aligned} \tag{3.3–4d}$$

(3.3–3c) goes with (3.3–4c) over into

$$(M_p)^2 = C_0 - \sum_{q=1}^{n_q} a_{pq}\, C_{pq} \ . \tag{3.3–4e}$$

If P_p is identified with a source point Q_n, then (3.3–4d, 3a) give the results

$$a_{nq} = \delta_{nq} = \begin{cases} 0 & \textit{for} \quad n \neq q \\ 1 & \textit{for} \quad n = q \end{cases} \qquad \Rightarrow \quad \hat{y}_n = y_n \ . \tag{3.3–4f}$$

The **covariance function** C_{pq} is selected as a function of distance

$$s_{pq} \ \ \textit{(distance between the points } Q_p \textit{ and } Q_q\textit{)} \, . \tag{3.3–5a}$$

Q_p, Q_q are the foot points of P_p, P_q in the reference surface of the surface normal coordinates. A representation of the covariance function that is in general well adapted to the specific requirements can be achieved with exponential functions:

$$\begin{aligned} C_{pq} &= c_1 \exp(-d_1\, s_{pq}^2) + c_2 \exp(-d_2\, s_{pq}^2) \ , \\ c_1 + c_2 &= C_0 \ , \quad d_2 \ll d_1 \end{aligned} \tag{3.3–5b}$$

(cf. Heitz 1968). The first exponential function describes the covariance for the near zone and the second one must ensure a sufficiently strong decrease in the distant zone in order to avoid singularity of the equation system (3.3–4c). This can in any case be achieved with a nonstationary covariance function only if (3.2–5a) is fulfilled:

- **Observation stations** (3.3–6a)
  or source points (3.1–2a) are to be selected nearly equidistantly.

(3.3–6a) can be achieved analogously to (3.2–5c) through a completion of the observations y_q by

- **linearly interpolated $\bar{y}_q$–values** (3.3–6b)
  for source points $\bar{P}_q$ between neighbouring observation stations P_q (triangle or quadrangle meshes)

These 'local linear interpolations' on the basis of three or four neighbouring values of observations can again be realized by means of collocation or polynomial representations.

To compute the internal and external deformation measures presented in *sections 2.3.2 and 2.3.3*, the first and second partial derivatives of $\hat{y}_p$ according to the surface coordinates u^α

$$\hat{y}_{p,\alpha} := (\partial\hat{y}/\partial u^\alpha)_p \ , \qquad \hat{y}_{p,\alpha\beta} := (\partial^2\hat{y}/\partial u^\alpha\, \partial u^\beta)_p \tag{3.3–7a}$$

are needed. Due to (3.3–1b, 4d) one obtains generally

$$\hat{y}_{p,\alpha\ldots} = \sum_{q=1}^{n_q} a_{pq,\alpha\ldots}\; y_q \qquad a_{pq,\alpha\ldots} = \sum_{q=1}^{n_q} C_{ql}^{-1}\; C_{pl,\alpha\ldots} \ . \tag{3.3–7b}$$

For the covariance function (3.3–5b), the partial derivations result to:

$$\begin{aligned} C_{pl,\alpha} &= -[\,c_1 d_1 \exp(-d_1\; s_{pq}^2) + c_2 d_2 \exp(-d_2\; s_{pq}^2)\,]\, s_{pl,\alpha}^2 \ , \\ C_{pl,\alpha\beta} &= [\,c_1 d_1^2 \exp(-d_1\; s_{pq}^2) + c_2 d_2^2 \exp(-d_2\; s_{pq}^2)\,]\, s_{pl,\alpha}^2\, s_{pl,\beta}^2 \\ &\quad -[\,c_1 d_1 \exp(-d_1\; s_{pq}^2) + c_2 d_2 \exp(-d_2\; s_{pq}^2)\,]\, s_{pl,\alpha\beta}^2 \ ; \\ s_{pl,\alpha}^2 &:= \partial s_{pl}^2/\partial u^\alpha \ , \qquad s_{pl,\alpha\beta}^2 := \partial^2 s_{pl}^2/\partial u^\alpha\, \partial u^\beta \ . \end{aligned} \tag{3.3–7c}$$

In Cartesian coordinates one finds

$$s_{pq}^2 = (x_{1.q} - x_{1.p})^2 + (x_{2.q} - x_{2.p})^2 \ , \tag{3.3–8a}$$

and therefore

$$\begin{aligned} s^2_{pq,1} &= -2\,(x_{1.q} - x_{1.p})\ , \qquad s^2_{pq,2} = -2\,(x_{2.q} - x_{2.p})\ , \\ s^2_{pq,11} &= 2\ , \quad s^2_{pq,12} = 0\ , \quad s^2_{pq,22} = 2\ . \end{aligned} \tag{3.3–8b}$$

In ellipsoidal coordinates one may start from the following approximations:

$$s^2_{pq} = R^2\,\{[(\lambda_q - \lambda_p)\cos\phi_m]^2 + (\phi_q - \phi_p)^2\}\ , \tag{3.3–9a}$$

$$\phi_m = (\phi_q + \phi_p)/2\ , \quad R = \textit{mean ellipsoidal radius}\ (R_1\,R_2)^{1/2}$$

and

$$\begin{aligned} s^2_{pq,1} &= -2R^2\,[(\lambda_q - \lambda_p)\cos^2\phi_m]\ , \quad s^2_{pq,2} = -2R^2\,(\phi_q - \phi_p)\ , \\ s^2_{pq,11} &= 2R^2\cos^2\phi_m\ , \qquad s^2_{pq,12} = 0\ , \\ s^2_{pq,22} &= 2R^2\ . \end{aligned} \tag{3.3–9b}$$

3.4 Subdivision of a Surface by Finite Triangle Elements

3.4.1 Preliminary Remarks

Seen from a general point of view, the **Finite Element Method** (FEM) constitutes a numerical method for solving initial boundary value problems of partial differential equations on the basis of a principle of variations. It has gained, among others, great importance for the solution of problems in continuum mechanics (see, e.g., Heitz 1980–1983; Zienkiewicz 1972). Here, only a partial characteristic of the FEM is needed, namely the geometric modelling of a surface in the form of a **subdivision by finite elements** or a **triangulation**, which means the spanning of a surface by means of small but finite partial surfaces, i.e. the so-called finite elements, which at their boundaries meet certain conditions of continuity. The simplest element is the plane or curved **triangle element**, for which an unambiguous linear interpolation of the corner or node values of the heights and displacements is possible. In the following only this triangulation is treated and applied, since besides its mathematical simplicity it has the outstanding feature that it can be applied particularly well directly to inhomogeneously distributed observation stations.

Two applicabilities of the triangulation of a surface are possible within the scope of the problems being discussed in this book, i.e. the determination of internal and external deformation measures of surfaces. On the one hand the subdivision by finite triangle elements may serve to create a rigorously or approximately equidistant control point net, on the basis of which interpolation by means of polynomials or collocation are possible according to the representations in *sections 3.2 and 3.3* [cf. in particular (3.2–5a) and (3.3–5b)]. In this way the subdivision by finite triangle elements continuously differentiable in pieces is transformed into a geometric modelling which is analytical in the whole area, so that the deformation measure discussed in *section 2.3* can be computed directly for any choice of points. For this purpose the foundations of the subdivision by finite triangle elements treated in the following *section 3.4.2* are sufficient. On the other hand one may dispense with analytical modelling for the whole area if on the basis of the subdivision by finite triangle elements suitable discrete deformation measures are defined, which will be dealt with in *section 3.4.3.*

3.4.2 Triangle Elements

The surface region F is decomposed into non–intersecting triangles whose corners coincide with observation stations P_q, (3.1–2a). Here, the designations and fundamentals introduced in *section 3.1*, also hold true, which are in each case to be applied to one

finite triangle Δ:

$$\textit{corners } P_q \ , \qquad q \in \{1,2,3\} \tag{3.4–1a}$$

For

$$\textit{test points } P_p \in \Delta \tag{3.4–1b}$$

a linear interpolation procedure (3.1–5a) is used:

$$\hat{y}_p = \sum_{q=1}^{3} a_{pq}\, y_q \ , \qquad \hat{y}_p := \hat{y}_p^T \ , \quad y_q := y_q - y_q^T \ , \tag{3.4–1c}$$

for which the collocation as well as the representation of polynomials are suited in principle. The latter is chosen because it can in this case be used easily and clearly. Within a finite triangle Δ the heights as well as the displacement coordinates (3.1–1b, c),

$$y(u^\alpha) \in \{H(u^\alpha)\ ,\ \ z^a(u^\alpha)\}\ , \tag{3.4–2a}$$

are determined unambiguously by a polynomial of the first degree:

$$\hat{H}(u^\beta) = c + c_\beta\, u^\beta\ , \qquad \hat{z}^a(u^\beta) = c^a + c^a_\beta\, u^\beta\ . \tag{3.4–2b}$$

The coefficients c , c_β and c^a , c^a_β, respectively, are computed via the conditional equations in the corners P_q

$$H_q = c + c_\beta\, (u^\beta)_q\ , \qquad (z^a)_q = c^a + c^a_\beta\, (u^\beta)_q\ , \tag{3.4–2c}$$

which may be applied analogously to the computations represented in Heitz (1980–1983) for three-dimensional tetrahedron elements. With the introduction of the *e–tensor*

$$e_{\alpha\beta} := \begin{vmatrix} \delta^1_\alpha & \delta^1_\beta \\ \delta^2_\alpha & \delta^2_\beta \end{vmatrix} \tag{3.4–3}$$

the following results are obtained:

$$\hat{H}_p = \sum_{q=1}^{3} a_{pq}\, H_q\ , \tag{3.4–4a}$$

$$(\hat{z}^b)_p = \sum_{q=1}^{3} (a^b_c)_{pq}\, (z^c)_q = \sum_{q=1}^{3} a_{pq}\, (z^b)_q \tag{3.4–4b}$$

with:

$$\begin{aligned} a_{pq} &= [b_q + (b_\beta)_q\, u^\beta]/D\ , \\ b_q &= -(-1)^q e_{\beta\gamma}\, (u^\beta)_{q+1}\, (u^\gamma)_{q+2}\ , \\ (b_\beta)_q &= (-1)^q e_{\beta\gamma}\, [(u^\gamma)_{q+2} - (u^\gamma)_{q+1}]\ , \\ & \quad \textit{if } q+n > 3\ ,\ \ \textit{then } q+n =: q+n-3\ , \\ D &= e_{\alpha\beta}\, [(u^\alpha)_2 - (u^\alpha)_1]\, [(u^\beta)_3 - (u^\beta)_1]\ . \end{aligned} \tag{3.4–4c}$$

Besides this representation the interpolation formulas (3.4–2b) that are explicit in the test point coordinates u^β are of interest in this context. For the coefficients one obtains with (3.4–4c):

$$c = \sum_{q=1}^{3}[b_q/D]\ H_q\ , \qquad c_\beta = \sum_{q=1}^{3}[(b_\beta)_q/D]\ H_q\ ,$$
$$c^a = \sum_{q=1}^{3}[b_q/D]\ (z^a)_q\ , \qquad c^a_\beta = \sum_{q=1}^{3}[(b_\beta)_q/D]\ (z^a)_q\ . \tag{3.4–5a}$$

The partial derivations of the heights and displacement coordinates according to the test point coordinates u^β result from (3.4–2b) with the mode of designation (3.3–7a) in:

$$\hat{H}_{,\alpha} = c_\alpha\ , \qquad \hat{H}_{,\alpha\beta} = 0\ ,$$
$$\hat{z}^a_{,\alpha} = c^a_\alpha\ , \qquad \hat{z}^a_{,\alpha\beta} = 0\ . \tag{3.4–5b}$$

The **boundary lines of the finite triangles** Δ are chosen such that in the reference surface of the surface normal coordinates linear equations

$$d + d_\beta\ u^\beta = 0 \tag{3.4–6}$$

apply to them, whose coefficients are determined by the coordinates $(u^\beta)_q$ of the corners P_q. The interpolation values (3.4–2b, 4a, b) are continuous at the 'triangle sides'.

As a rule, the **triangle surfaces are curved**, but if Cartesian coordinates are used according to (2.1–4) they are planes, the triangle sides in this case being straight lines.

3.4.3 Discrete Deformation Measures

Taking into account (3.4–5b), the **linear surface deformation tensor** according to (2.3–10b) can be directly computed for a finite triangle Δ (3.4–1):

$$\begin{aligned}\varepsilon_{\alpha\beta} \;=\; &\{\, q^c_{,\alpha}\,[c^d_\beta + \Gamma^d_{ef}\, q^e_{,\beta}\,(c^f + c^f_\gamma\, u^\gamma)] \\ &+ q^c_{,\beta}\,[c^d_\alpha + \Gamma^d_{ef}\, q^e_{,\alpha}\,(c^f + c^f_\gamma\, u^\gamma)]\,\}\, g_{cd}/2\end{aligned} \tag{3.4–7a}$$

and the **infinitesimal rotation vector** according to (2.3–36b):

$$\tilde{d}^c = (1/2)\, \tilde{\varepsilon}^{\alpha bc}\, q^g_{,\tilde{b}}\, [\, c^d_{\tilde{\alpha}} + \Gamma^d_{ef}\, q^e_{,\tilde{\alpha}}\, (c^f + c^f_\gamma\, u^\gamma)\,]\, g_{dg} \; . \qquad (3.4\text{–}7b)$$

Both results depend in general on the selection of the

$$\textit{test point } P_p(u^\gamma) \in \Delta \qquad (3.4\text{–}8a)$$

Only in cases that Cartesian coordinates are selected as external surface normal coordinates, in which the triangle surfaces are planes, this dependence does not apply.

If the sides of the triangle Δ are small as compared to the radii of curvature of the reference surface, then the dependencies on (3.4–7a, b) are also small. In this case it is in general sufficient to select a mean test point P_Δ with the coordinates

$$(u^\gamma)_\Delta := (1/3) \sum_{q=1}^{3} (u^\gamma)_q \; ,$$
$$P_q = \textit{corner of the finite triangle } \Delta. \qquad (3.4\text{–}8b)$$

As **internal surface deformation measures** are appropriately computed:

$$\bar{q}_\Delta \;=\; f^{\alpha\beta}\, \varepsilon_{\alpha\beta} \qquad (3.4\text{–}9a)$$
$$\;=\; \textit{surface dilatation in } P_\Delta \textit{ according to (2.3–7b)} \; ,$$

$$q_{(n)\Delta} \;=\; \varepsilon_{\alpha\beta}\, (r^\alpha)_n\, (r^\beta)_n \; , \qquad n \in \{1,\, 2\} \qquad (3.4\text{–}9b)$$
$$\;=\; \textit{extreme values of the linear elongation in } P_\Delta \textit{ according to (2.3–7g)} \; ,$$

$$\varepsilon_{\alpha\beta} \qquad \textit{according to (3.4–7a)} \; .$$

To represent the external deformations of the surface, the changes of surface curvatures mentioned in section 2.3.3.2 do not enter into consideration with the subdivision of a surface by finite elements. To judge the **changes of the external surface geometry,** the infinitesimal rotation vectors $(\tilde{d}^c)_\Delta$ calculated for the mean test points P_Δ according to (3.4–7b) can be used, if they are transformed to the local bases of the corners P_q of the triangles Δ:

$$(d^e)_q \;=\; (q^e_{,\tilde{c}})_q\, (\tilde{d}^c)_\Delta \quad \textit{for} \quad q \in \{1,2,3\} \qquad (3.4\text{–}10)$$
$$\;=\; \textit{mean rotation vector of } \Delta \textit{ transformed to } P_q \; .$$

In the local bases of the corners P_q the rotation vectors $(d^e)_{q(n)}$ of the adjoining triangles $\Delta_{(n)}$, $n = 1,2,3,4\ldots$ are comparable, and from their differences external deformation data can be derived. As another possibility for the description of the changes of the external surface geometry, the change of the bends between the neighbouring triangles can be used.

3.4.4 Determination of the Change of the Bend Between the Neighbouring Triangles

After spanning the investigation area in the triangles by the selection of three noncollineatory points, for each triangle the *normal unit vector* n_i, which is perpendicular to each point in the triangle, can be determined (*Fig. 3.1*):

$$n_i = \frac{\varepsilon_{ijk}\, x_{j.12}\, x_{k.13}}{|\varepsilon_{ijk}\, x_{j.12}\, x_{k.13}|} \qquad (3.4\text{–}11a)$$

with $x_{i.12} = x_{i.2} - x_{i.1}$ und $x_{i.13} = x_{i.3} - x_{i.1}$.

The relation between the normal unit vectors of two neighbouring triangles D_1 and D_2 is given with

the *cosine of the angle*

$$\cos\alpha = \left(\frac{n_{i.D_1}(t)\, n_{i.D_2}(t)}{|n_{i.D_1}(t)|\,|n_{i.D_2}(t)|}\right) \qquad (3.4\text{–}11b)$$

or with

the *sine of the angle*

$$\sin\alpha = \left(\frac{|\varepsilon_{ijk}\, n_{j.D1}\, n_{k.D2}|}{|n_{i.D_1}|\,|n_{i.D_2}|}\right), \qquad (3.4\text{–}11c)$$

where the unit vectors are $|n_{i.D_1}| = |n_{i.D_2}| = 1$.

According to (3.4–11b, c) the change of the bend between two neighbouring triangles for the observation time t and $\bar{t}$ results to (*Fig. 3.2*)

$$\Delta\alpha_{12} = \alpha_{12}(\bar{t}) - \alpha_{12}(t)\ . \qquad (3.4\text{–}12)$$

If the change of the bends for all neighbouring triangles are determined, the changes of the external form of the investigation area can be illustrated.

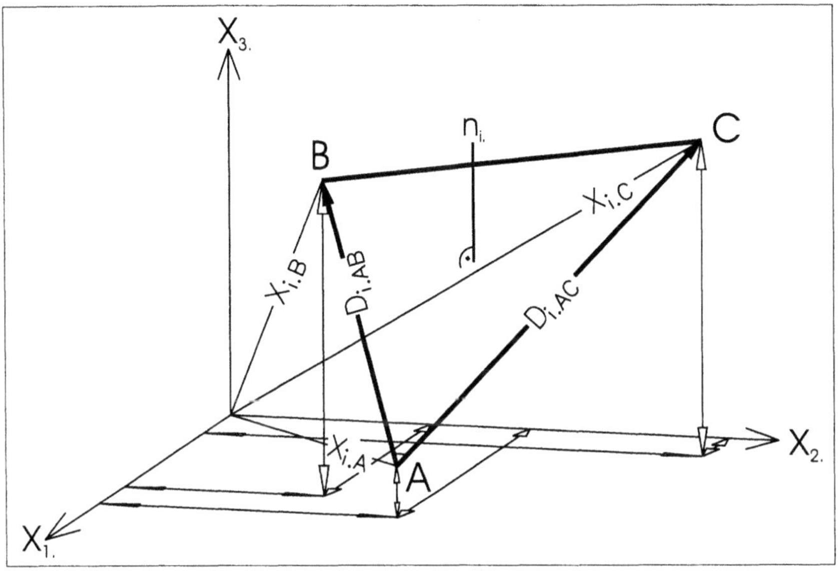

Fig. 3.1. The normal unit vector n_i which is perpendicular to each point in the triangle

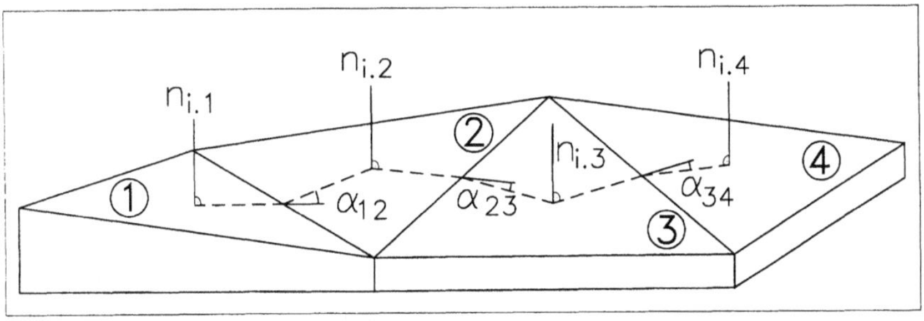

Fig. 3.2. The change of the bend between the triangles describes the changes of the external form of the investigation area

Chapter 4

Application

4.1 Deformation of the Earth's Crust

All points of the Earth's crust are subject to deformation. The most important causes for the permanent and temporal displacements of points on the Earth's crust are plate tectonic motion, tidal effects, atmospheric, hydrological, ocean loading, and local geological processes (see, e.g., Melchior 1983; Heitz 1980–1988). For precise geodetic applications of the satellite techniques, plate motion and tidal effects are to be taken into consideration for the definition of the absolute station coordinates of a reference frame as well as in the modelling of orbits, at least to the extent that the network is located on a simple plate. The remaining effects induce short–term temporal changes of the station coordinates, influences which for a small investigation area can be reduced by using a long time span for GPS observations, e.g. 24 hours observations for each session.

The realizations of the International Terrestrial Reference Frame (ITRF) as well as the European Terrestrial Reference Frame 1989 (ETRF89) are accomplished for precise geodetic applications (see, e.g., Boucher et al. 1992, 1993, 1994; McCarthy 1996; Altıner et al. 1995, 1997a, 1997b; Mišković and Altıner 1997; Seeger et al. 1998). The World Geodetic System 1984 (WGS 84) agrees with the ITRF and ETRF89. The differences between the WGS 84 and the ITRF are in the cm–range worldwide (see NIMA 1997). For control and promotion of such a reference frame, continuous observations through permanent stations distributed at a worldwide level are necessary (*Fig. 4.1*). The-

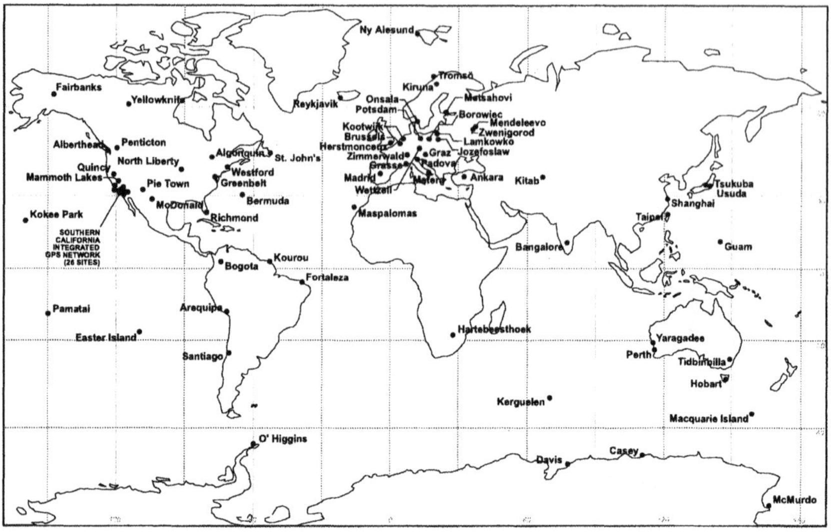

Fig. 4.1. Tracking IGS stations (Schneider 1998)

se continuously performed observations on permanent stations allow us to determine plate motion precisely, to compute the Earth rotation parameters, and to investigate the reaction of the Earth as a deformable body due to the attraction by the Moon and the Sun.

From the sixties onwards the movement of the Earth's lithospheric plates has been described on the basis of the analysis of global ocean floor spreading rates, transform fault systems, and earthquake slip vectors (see, e.g., Kent 1976; Miller 1992). According to this theory, the Earth's crust consists of 14 to 16 major lithospheric plates, floating on the fluid asthenosphere (*Fig. 4.2*). On the basis of the point of action the forces moving the lithospheric plates can be classified into two groups. One group consists of those forces acting on the whole plate. These forces arise from convection cells in the upper mantle. The second group consists of the forces that act on the boundaries of the plates. At the mid–oceanic ridge hot magmatic material emerges, cools and forms new rocks, spreading the ocean floor apart. By this spreading the plates are shifted on their boundaries and begin to move. The forces subducting the plates also belong to the second group. As a result of these motions, energy is released, the magnitude of which depends on the size and thickness of the plates and on the rate of motion.

Some models dealing with global plate motions were published in (DeMets et al. 1990, 1994; Argus and Gordon 1991; Seno et al. 1987, 1983; Wilson 1993a, 1993b). These models describe plate motions by a rotation vector (Euler Vector) of the plate based on a pole of rotation. The NUVEL–1 model describes motions between 14 major rigid plates relative to the fixed Pacific plate (Demets et al. 1990). The NNR–NUVEL1 model (no net rotation) gives absolute angular velocities of the plates (Argus and Gordon 1991). Due to recent revision in the paleomagnetic time scale, the angular velocities of these earlier models are multiplied by a recalibration factor of 0.9562 (DeMets et al. 1994). These rescaled models are known as NUVEL–1A and NNR–NUVEL1A global models. As an example, the angular velocities of the lithospheric plates in the NUVEL–1A model are given in *Table 4.1* (Demets et al. 1994).

Plate	Lat. ° [N]	Long. ° [E]	ω °/Myr	ω_x rad/Myr	ω_y rad/Myr	ω_z rad/Myr
Africa	59.160	-73.174	0.9270	.002400	-.007939	.013892
Antarctica	64.315	-83.984	0.8695	.000689	-.006541	.013676
Arabia	59.658	-33.193	1.1107	.008195	-.005361	.016730
Australia	60.080	1.742	1.0744	.009349	.000284	.016252
Caribbean	54.195	-80.802	0.8160	.001332	-.008225	.011551
Cocos	36.823	251.371	1.9975	-.008915	-.026445	.020895
Eurasia	61.066	-85.819	0.8591	.000529	-.007235	.013123
India	60.494	-30.403	1.1034	.008180	-.004800	.016760
North America	48.709	-78.167	0.7486	.001768	-.008439	.009817
South America	54.999	-85.752	0.6365	.000472	-.006355	.009100
Nazca	55.578	-90.096	1.3599	-.000022	-.013417	.019579
Rivera	31.000	257.600	2.4500	-.007880	-.035800	.022020
Scotia	49.100	-89.400	0.6600	.001100	-.007500	.008700
Juan de Fuca[1]	35.000	26.000	0.5100	.006510	.003170	.005080
Juan de Fuca[2]	28.300	29.300	0.5200	.006710	.003770	.004150
Philippine[3]	0.000	-47.000	0.9600	.011400	-.012200	.000000
Philippine[4]	-1.200	-45.800	0.9600	.011600	-.012000	.000300

Table 4.1. Angular velocities of the NUVEL–1A model (DeMets et al. 1994): *1) The Juan de Fuca plate is recalibrated from Wilson (Wilson 1993a); 2) The Juan de Fuca plate is recalibrated from the more recent estimate of Wilson (Wilson 1993b); 3) The Philippine plate is recalibrated from Seno (Seno et al. 1987); 4) The Philippine plate is recalibrated from the more recent estimate of Seno (Seno et al. 1993)*

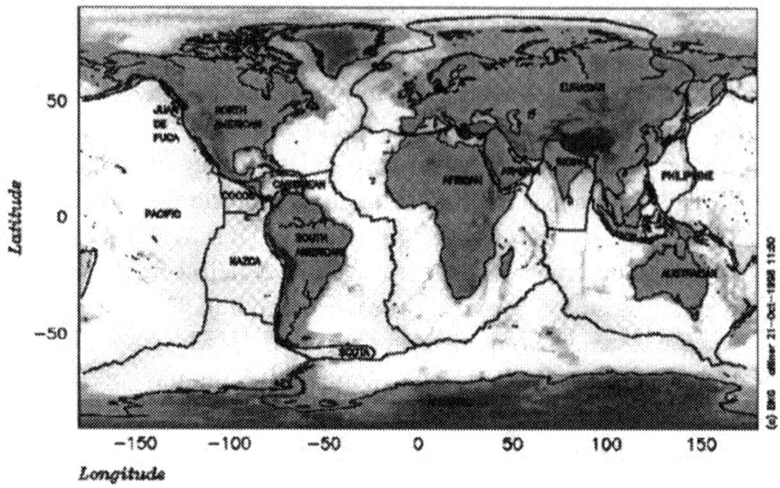

Fig. 4.2. Major tectonic plates and plate boundaries

On a spherical Earth the velocities of a station on a plate are given by

$$z_{i.s} = R\,\omega_p \begin{bmatrix} \cos\phi_p \sin\lambda_p \sin\phi_s \; - \; \sin\phi_p \cos\phi_s \sin\lambda_s \\ \cos\phi_s \cos\phi_s \sin\lambda_s \; - \; \cos\phi_p \sin\phi_s \cos\lambda_p \\ \cos\phi_p \cos\phi_s \sin(\lambda_s - \lambda_p) \end{bmatrix} , \qquad (4.1\text{–}1)$$

where $(\lambda,\ \phi)_s$ are the coordinates of the station on the plate, $(\lambda,\ \phi)_p$ are the coordinates of the rotation pole of the plate, and R is the spherical Earth's radius. ω describes the angular velocity of the plate.

With

$$x_{i.s}(t) \;=\; x_{i.s}(t_0) \;+\; z_{i.s}(t - t_0) \qquad (4.1\text{–}2)$$

one obtains for the observation epoch t the corrections of the station's coordinates due to the global plate motion relative to a reference epoch t_0. Gravitational attraction of the Moon and the Sun on points on the Earth's crust, illustrated in (*Fig. 4.3*), is not constant because the Moon and the Sun change position due to the rotation of the Earth. This changing attraction of the Moon and the Sun is the main cause of oceanic and solid Earth tides. For the modelling of the tidal effects (see, e.g., Melchior 1983; Heitz 1980–1988; Vaniček and Krakiwski (1986); Sovers and Jacobs 1994; McCarthy 1996).

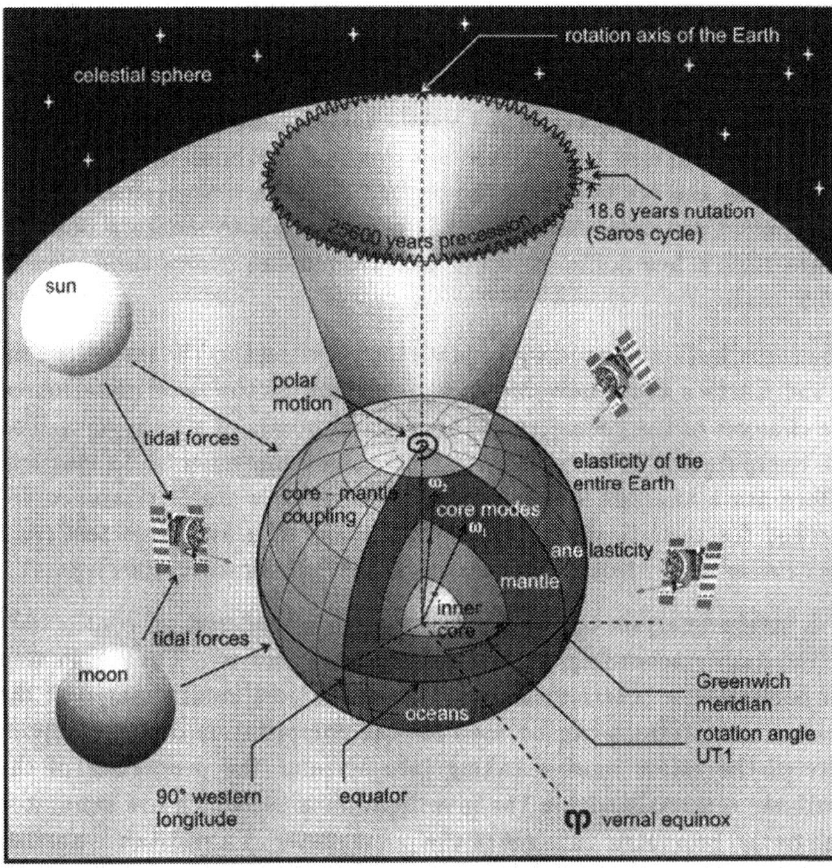

Fig. 4.3. Tidal effects cause temporal and permanent displacement of the stations (Schneider 1998)

4.2 Geodetic Contributions to Deformation Analyses

The horizontal motions of the litospheric plates generally range from a few to more than 150 mm/yr; smaller motions are expected in the vertical direction. The measurement of such small relative movements between points far away from each other has been made possible by the development of the space techniques (VLBI/Very Long Baseline Interferometry, SLR/Satellite Laser Ranging, and GPS). Such measurement could not be achieved by classical methods over the plate boundaries and faults due to the limited distance measurement of the electronical distance observers. It is now possible to measure motions of less than a few millimetres per year from space over a time span of a few years.

The contribution of geodesy to plate tectonics and to the investigation of the Earth's crust movements can be seen in the determination of the changes of the geometry of the selected points of a body as well as the computation of forces necessary for these movements. In classical deformation analysis, displacements relative to a stable reference are checked for statistical significance by means of a hypotesis test (see, e.g., Pelzer 1971; Koch 1985; Caspary 1987; Hekimoğlu 1997).

In a strain analysis the displacements are considered as continously differentiable acoording to the surface coordiates u^α. The strain tensor components determined by means of the positional changes of the observation stations can be used for the computation of the components of the stress tensors taking into account the properties of the available materials within the investigation area, which are generated due to the geometric changes of the point raster as a resistance against the force within the body that acts against the body. Therefore, the strain analysis can be considered as a basis of a dynamic model, whereas the classical deformation analysis is similar to a kinematic model (see, e.g., Flügge 1972; Means 1976; Grafarend 1977; Brunner 1979; Bock 1982; Welsch 1989; Ghitău 1998). *Chapter 4* has been dedicated to the applications of the geometric fundamentals of the analytical surface deformation theory elaborated in *chapters 2 and 3*. For the application, the coordinate differences in a mean epoch between the CRODYN'94 and CRODYN'96 campaigns, were used (ITRF94, epoch

1995.6). Internal and external deformation measures were determined for the description of the surface deformations within the area covered by the GPS network, which is established in 1994 in the surroundings of the Adriatic Sea area. Only these results are the subject of the examinations described in *sections 4.4 and 4.5.*

4.3 Tectonic Development of the Adriatic Sea Area

Seismological and geological studies indicate that the tectonic development of the Adriatic Sea area is dominated by the collision of the African plate with the Eurasian plate. The anticlockwise rotation of the African plate causes a northward movement of the northern part against the southern flank of the Eurasian plate. This northward movement of the African plate causes a great change in the tectonic evolution of the Eastern Mediterranean, giving rise to the Aegean extensional regime and it is the main cause of the eartquakes in the Adriatic Sea area (McKenzie 1972; Udias 1982; Anderson and Jackson 1987; Geiss 1987; Jackson and McKenzie 1988; Mantovani et al. 1992; Drewes 1993).

From the sixties onwards, this area was focused on by many scientists for the investigation of the present tectonic features. The investigation of the seismicity of the Mediterranean, the reinterpretation of the results evaluated, especially with the addition of the new knowledge about the latest earthquakes, were the basis for many publications dealing with the tectonics in this area. The results of these investigations can generally be summarized in two different directions. A group of scientists assume that the Adriatic Sea is a promontory of the African Plate (Channel 1979; D'argio and Horvath 1984; Mantovani et al. 1992; Babbucci et al. 1997). Mantovani et al. (1992) assert that this promontory moves in a northnorthwest direction with a magnitude of approx. 5 mm/yr (*Fig. 4.4, page 64*). Another group of the scientists describes the Adriatic Sea as an independent microplate and they refer this assumption to the concentration of the earthquakes in the land area, *see Fig. 4.5, page 65*, as well as to the recent tectonic style and crustal structure of this area (Celet 1977; Giese and Reutter

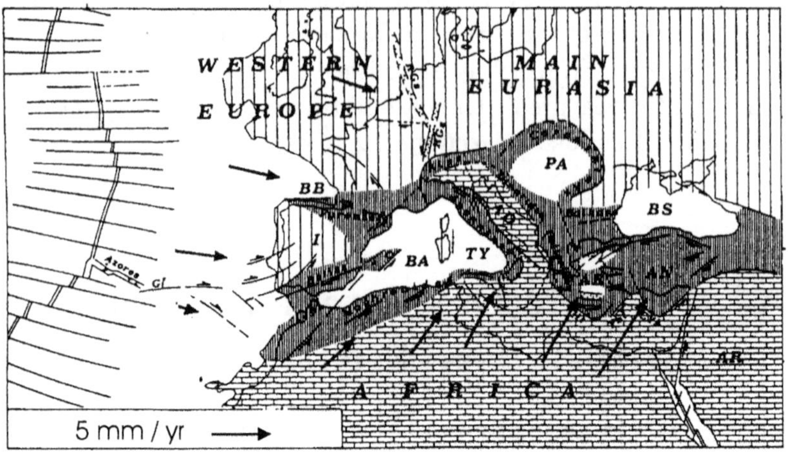

Fig. 4.4. The Adriatic Sea area is a promontory of the African plate and movement of this promontory in a north–northwest direction is about 5 mm/yr. AR: Arabia; AN: Anatolia; BS: Black Sea; AE: Aegea; AD: Adriatic block; PA: Pannonian basin; TY: Tyrrhenian basin; BA: Balearic basin; BB: Bay of Biscay; GM: Gibraltor–Morocco block; RGs: Rhine Graben system (Mantovani et al. 1992)

1978; Vandenberg and Zijderveld 1982, Anderson and Jackson 1987). Anderson and Jackson (1987) explains the earthquakes covered by the sea area as an internal deformation of the Adriatic microplate.

4.4 Geometric Modelling

4.4.1 Interpolation Methods Applied

The selection and observation of stations for the analysis of deformations of the Earth's surface are generally subject to constraints which in part lead to significant deviations from the requirements founded theoretically. This is also the case with projects treated here. Due partly to difficult topographic conditions it was not possible with the funds available to achieve a sufficiently homogeneous distribution of stations in the areas to be investigated. Therefore, for the observation area, height and displacement values were computed by interpolation in sufficiently dense, homogeneous grids. To do this the collocation method,

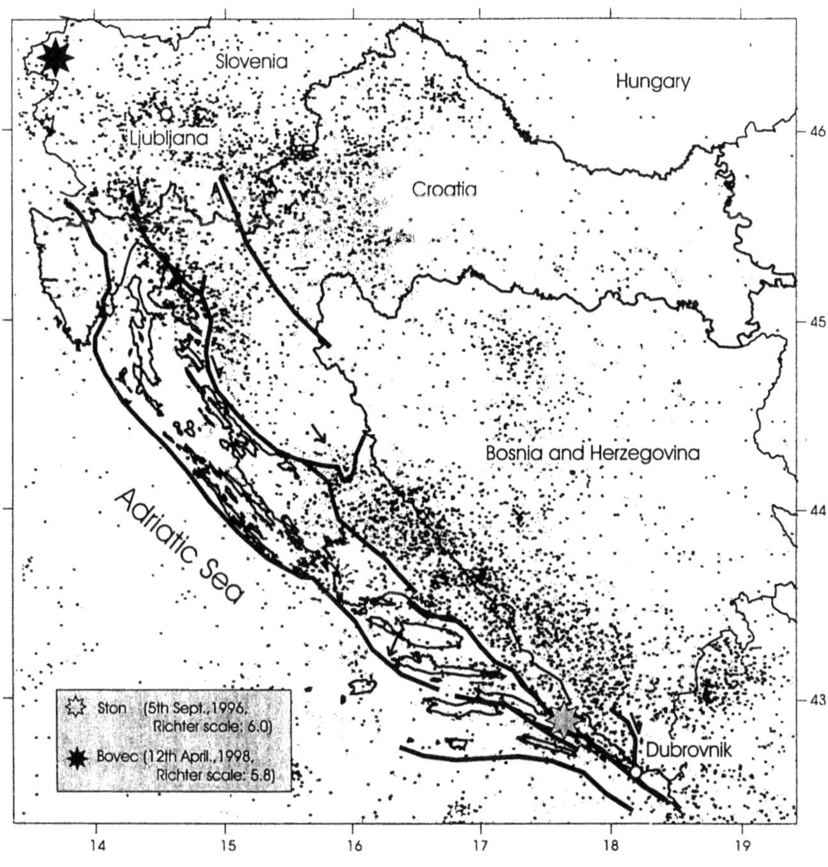

Fig. 4.5. Earthquakes occurred in the last 500 years in the Adriatic Sea area (Skoko and Mokrovic 1998)

as described in *sections 3.1 and 3.3*, was used. The heights and velocities were also interpolated by the spline method for the comparison of both results. In this case, the differences between the coordinates interpolated by the spline and collocation methods were very small and can be disregarded.

The method of the subdivision by finite triangle elements or triangulation is well suited for the computation of internal deformation measures [cf. (3.4–7a)], as well as for the densification of observation stations, because in this case only the functions themselves are continuous at the triangle sides, but not their derivatives. However, for the determination of the external deformation measures, the collocation method can be applied in a considerably simpler and clearer way, since it provides interpolation functions that can be differentiated in the whole area of interpolation two times continuously according to the surface coordinates u^α. Their application then allows us to compute directly all internal and external deformation measures on the basis of the formulas presented in *sections 2.3.2 and 2.4.3.*

Computation of the internal deformation measures was performed on the basis of the formulas (2.3–4a–10d) from *sections 2.3.2.2 and 2.3.2.3* taking into account the ellipsoidal coordinates [cf.*section 2.3.2.4*]. In contrast to this, the formulas (2.3–17a–37c) from *sections 2.3.3.1, 2.3.3.2 and 2.3.3.3* were used to compute the external deformation measures.

4.4.2 Explanations of the Graphic Representations

Here, the general fundamentals for the graphic representations of the following sections are discussed. The observation values and deformation measures for the Earth's surface F to be demonstrated, present the following characteristics:

- **Single data on the reference ellipsoid:**

  one value is to be represented per point

$$Z = Z(\lambda,\ \phi)\ . \qquad (4.4\text{–}1a)$$

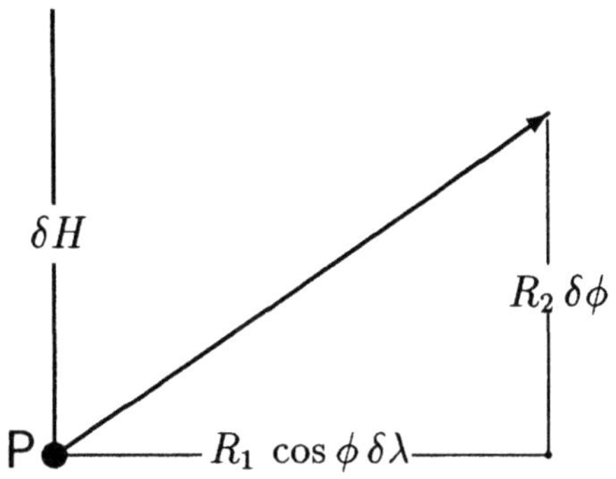

Fig. 4.6. Representation of displacement coordinates

- **Vector data on the reference ellipsoid:**

  three values are to be represented per point

$$Z^a = Z^a(\lambda,\ \phi)\ , \qquad a \in \{1,2,3\}\ . \tag{4.4–1b}$$

- **Vector data in the tangent plane of F:**

  two or more pairs of values are to be represented per point

$$Z^\beta_{(n)} = Z^\beta_{(n)}(\lambda,\ \phi)\ , \qquad \beta \in \{1,2\}\ , \quad n \in \{1,2\ldots\}\ . \tag{4.4–1c}$$

The **single data representations** (4.4–1a) comprise in particular *height representations* of the Earth's surface F (2.3–11a, b)

$$H = H(\lambda,\ \phi) \tag{4.4–2a}$$

as well as representations of the *surface dilatation* (2.3–7b) and of the *changes of the mean curvature* (2.3–21c)

$$\bar{q} = \bar{q}(\lambda,\ \phi)\ , \qquad \delta H = \delta H(\lambda,\ \phi)\ . \tag{4.4–2b}$$

For this **isoline representation** as well as **point grids with indication of values,** are suited.

The visualized *displacement coordinates* (2.3–11c),

$$\delta q^a(\lambda,\ \phi) = [\,\delta\lambda(\lambda,\ \phi),\ \delta\phi(\lambda,\ \phi),\ \delta H(\lambda,\ \phi)\,]^a\ , \tag{4.4–3a}$$

constitute **vector representations** (4.4–1b). A possibility is three separate **isoline representations** for

$$\delta\lambda(\lambda,\ \phi)\ , \qquad \delta\phi(\lambda,\ \phi)\ , \qquad \delta H(\lambda,\ \phi)\ . \tag{4.4–3b}$$

On the other hand **graphic vector representations** are also suited for this according to *Fig. 4.6.* The metric values

$$R_1 \cos\phi \cdot \delta\lambda(\lambda,\ \phi)\ , \qquad R_2 \cdot \delta\phi(\lambda,\ \phi) \tag{4.4–3c}$$

of the horizontal components in the form of two–dimensional vectors are appropriately traced in the drawing on map plane and the height component

$$\delta H(\lambda,\ \phi) \tag{4.4–3d}$$

is drawn in a line representation in the direction of ϕ– positively upwards and negatively downwards.

In the cases (4.4–1c), that is, with the **representation of tangent vectors,** it is appropriate to perform a visualization in the tangent planes in order to avoid distortions. An example to this is the *distortion vectors* (2.3–7h) in the tangential planes of F

$$m^{\alpha}_{(n)}(\lambda,\ \phi) = m_{(n)}\, r^{\alpha}_{(n)}\ , \quad m_{(n)} = 1 + q_{(n)}\ , \tag{4.4–4a}$$

which can be represented according to *Fig. 4.7.* With this $q_{(n)}$ is, as a rule, to be marked off in an enlarged form starting from the unit circle additionally represented

$$m_{(n)} = 1\ :\ \textit{no distortion.} \tag{4.4–4b}$$

Thus, with the enlargement factor $v > 1$ constitutes

$$m_{(n)} =: m_{(n)v} = 1 + v \cdot q_{(n)} \tag{4.4–4c}$$

instead of (4.4–4a) the line length in the drawing. Further, the tangent vector to the contour in the test point $P(\lambda,\ \phi,\ H)$

$$t^a = (\, t^1,\ t^2,\ 0\,)^a \tag{4.4–4d}$$

is indicated in the illustration. It is parallel to the reference ellipsoid and thus also to the basic plane of the representation so that it can be entered in the real direction. The computations required for this are explained in the following:

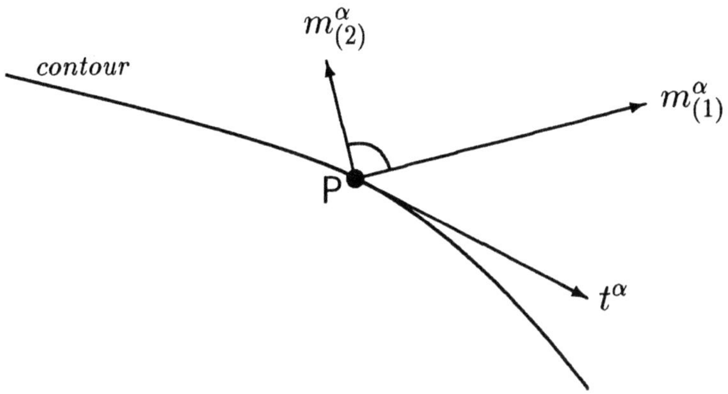

Fig. 4.7. Representation of distortion vectors

t^a is appropriately selected as unit vector so that the metric tensor (2.3–12a) applies

$$g_{11}\,(t^1)^2 + g_{22}\,(t^2)^2 = 1\;. \tag{4.4–5a}$$

Moreover, for the tangent vector to the contour the condition

$$H_{,\alpha}\,t^{\alpha} = H_{,\lambda}\,t^1 + H_{,\phi}\,t^2 = 0 \tag{4.4–5b}$$

must be met. By these two equations (4.4–5a, b), the components

$$t^1 \;=\; 1\,/\,(g_{11} + g_{22}\,v^2)^{1/2} \tag{4.4–5c}$$

and

$$t^2 \;=\; v\,t^1 \tag{4.4–5d}$$

with $v \;=\; -(H_{,\lambda}\,/\,H_{,\phi})$ can be determined. If $\bar{t}^{\alpha}$ are the components in the tangent plane, one obtains on the basis of the transformation equations

$$t^b = q^b_{,\alpha}\,\bar{t}^{\alpha} \tag{4.4–5e}$$

the relations

$$t^{\alpha} = \bar{t}^{\alpha}\;, \qquad t^3 = H_{,\lambda}\,\bar{t}^1 + H_{,\phi}\,\bar{t}^2 = H_{,\lambda}\,t^1 + H_{,\phi}\,t^2 = 0\;, \tag{4.4–5f}$$

the second equation in accordance with (4.4–5b). For the representation of the vectors (4.4–4a) in the tangent plane their components in the $\tilde{t}^\alpha$ direction and those perpendicular to this direction are needed, which can be computed as follows:

$$\tilde{m}^1_{(n)} = f_{\alpha\beta}\, m^\alpha_{(n)}\, t^\beta \;, \qquad \tilde{m}^2_{(n)} = f_{\alpha\beta}\, m^\alpha_{(n)}\, \varepsilon^\beta_\gamma\, t^\gamma \tag{4.4–5g}$$

with the metric tensor $f_{\alpha\beta}$ according to (2.3–16). It is

$$\varepsilon^\beta_\gamma = (f_{\gamma 1}\, \delta^\beta_2 - f_{\gamma 2}\, \delta^\beta_1)\, f^{-1/2} \tag{4.4–5h}$$

the two-dimensional ε–tensor (cf. Heitz 1988).

To visualize the *changes of the principal curvatures* (2.3–21b):

$$\delta\kappa_{N(n)} = \bar{\kappa}_{N(n)} - \kappa_{N(n)} \;, \tag{4.4–6a}$$

the vectors

$$k^\alpha_{N(n)}(\lambda,\, \phi) = \delta\kappa_{N(n)}\, r^\alpha_{(n)} \tag{4.4–6b}$$

are appropriately represented. For this purpose only the original direction vectors $r^\alpha_{(n)}$ are used; a representation of the directions of mean curvatures $\bar{r}^\alpha_{(n)}$ does not seem appropriate. In contrast to the positively defined distance conditions the changes of curvature may be positive and negative.

4.5 Surface Deformations in the Adriatic Sea Area

4.5.1 GPS Observations

The global plate tectonic models, described in *sections 4.1*, were developed on the basis of the ocean floor spreading rates, transform fault azimuths and earthquakes slip vectors which obtained generally on the plate boundary zones. Motions given in these models are an average of the plate movements in a time span over the past few million years (Myr). Therefore, the investigation of the present tectonic activities by the global plate motion models in an area as Adriatic Sea with complex movements can not give satisfactory results. For that reason, to study the present-day deformations in the Adriatic Sea area a GPS network was established in 1994 in co-operation among the Bundesamt für Kartographie und Geodäsie, Frankfurt am Main, the Geodetic Faculty of the University of Zagreb, the State Geodetic Administration in Zagreb and the Surveying and Mapping Authority in Ljubljana. The network consists of 22 stations which are distributed over Croatia (17), Slovenia (3) and Italy (2). The average distance between the points in the network amounts to 30 km. In 1996, the network was extended towards the south and west with 7 new stations (3 in Italy and 4 in Albania) in order to focus on present tectonic activities within the sea area (*Fig. 4.8, page 72*).

Within the CRODYN'94 campaign the first GPS observations in this network were carried out between June 7 and June 10, 1994 in three session with a session length of 24 hours (*Table 4.2, page 71*). The data were collected with an elevation cut-off angle of 15 degrees and a sampling rate of 15 seconds (Altıner et al. 1995). Because of variations in the phase center of different receiver antennas, all stations, except for two stations in Italy (Bassoviza and Trieste, Leica SR 299 receivers), were occupied with Trimble 4000 SSE and SSI receivers (see, e.g., Wübbena et al. 1997; Kaniuth et al. 1998). The second campaign in this network were conducted from September 9 to September 12, 1996 within the CRODYN'96 campaign with the same strategy as in 1994 (Altıner et al. 1997b).

Year	Day	Duration of observations		Sampling
	from – to	number of sessions	duration of a session from (UT) – to (UT)	rate in seconds
1994	158 – 160	3	9:00 – 9:00	15
1996	253 – 255	3	7:00 – 7:00	15
1998	247 – 249	3	8:00 – 8:00	15

Table 4.2. Duration of observations and average number of received satellites

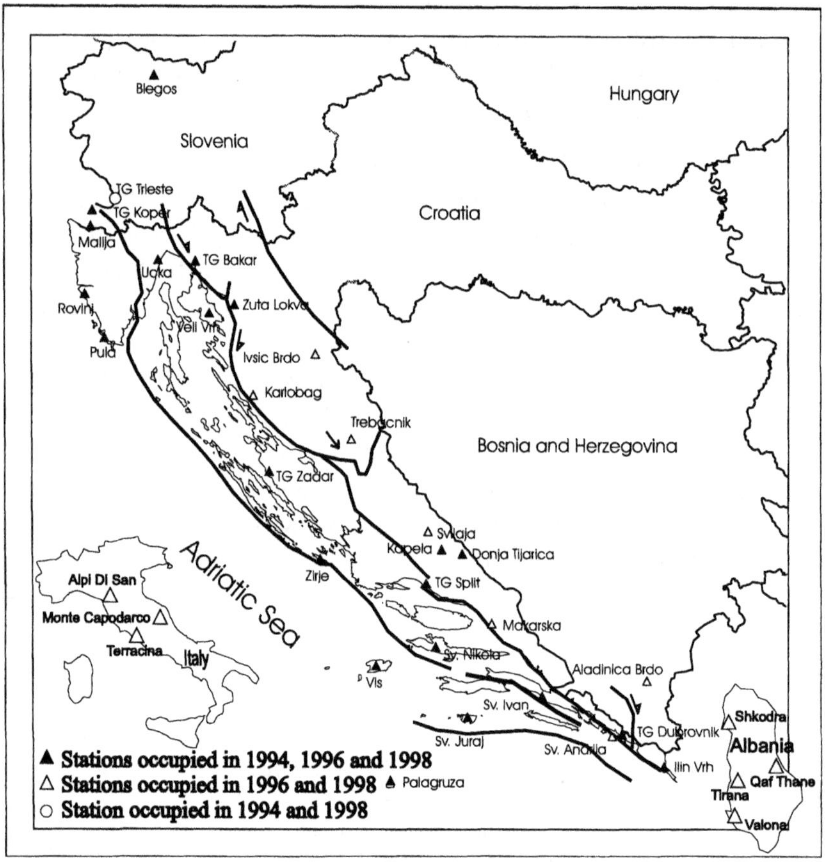

Fig. 4.8. GPS network in the Adriatic Sea area

4.5.2 Processing Strategy of the GPS Data

The processing of the data obtained by the CRODYN'94 and CRODYN'96 campaigns was completed using the Bernese GPS Software (Rothacher and Mervart 1996). The CODE orbits (Center for Orbit Determination in Europe) were used as precise orbit in the processing (*Table 4.3*).

Campaign	Orbit Reference Frame
CRODYN'94	ITRF92, epoch 1994.7
CRODYN'96	ITRF94, epoch 1996.7

Table 4.3. The individual reference frames of the CODE orbits used in the processing

The coordinates of the IGS sites (International GPS Service for Geodynamics) Graz (Austria), Matera (Italy), Zimmerwald (Switzerland) and Wettzell (Germany), which are included in the processing to define the datum of the reference frames, were shifted to the individual observation epochs of the CRODYN'94 and CRODYN'96 campaigns using the coordinates and velocity values published in the IERS Technical Notes for the ITRF92 and ITRF94.

4.5.3 Data Processing

In a first step of the data processing the obtained raw data were converted from the receiver format into the Receiver Independent Exchange (RINEX) format and antenna heights measured in the field were corrected on the basis of the $L1/L2$ antenna phase center of the antenna types used (Gurtner et al. 1989). In a second step, using pseudorange data the receiver clock correction for each observation epoch were performed by the single point positioning solutions and then stored in the phase zero difference files. After that, single differences for phase data were performed using the *shortest baseline* strategy. After forming single differences, cycle slips were removed using the $L3$ linear combination of $L1$ and $L2$ phase measurements in a baseline by baseline mode. In the same step, a priori coordinates were improved by the triple difference screening. With these a priori coordinates a free campaign solution (no ambiguity resolution) was conducted to check the processing strategy and to improve the a priori station coordinates (except the coordinates of the IGS stations used in the processing).

After that, the $L1$ and $L2$ ambiguities were resolved using the *Sigma Ambiguity Resolution* strategy of the Bernese software for the very short baselines and the *Quasi–Ionosphere–Free Ambiguity Resolution* strategy for the other baselines. About 95% of the initial $L1$ and $L2$ phase ambiguities were resolved successfully.

4.5.4 Checking of Data Quality

In order to check the data and processing quality daily solutions for each individual campaign were determined separately. For the computation of the daily solutions a priori uncertainty (standard deviation) of 1 m was introduced for the station coordinates (free network solution). The resolved $L1$ and $L2$ phase ambiguities were taken into consideration as known parameters and pre–eliminated before the inversion of the normal equations was computed. Zenith delay parameters for each two hours were computed by means of the Saastamoinen standard model (Saastamoinen 1972). The general characteristics of the adjustments for the daily solutions within the CRODYN'94 and CRODYN'96 campaigns are given in *Table 4.4 and 4.5.*

CRODYN'94 Day	RMS (in M) [Single Difference]	No. of Observations	No. of Parameters	No. of Coordinates
158	0.003	532566	359	69
159	0.003	501872	378	75
160	0.003	423729	375	72

Table 4.4. Characteristics of the adjustment for the daily solutions within the CRODYN'94

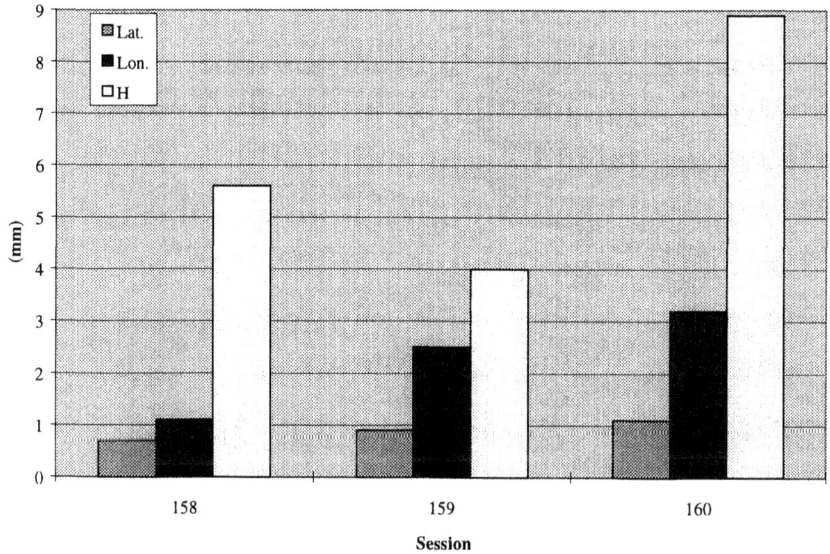

Fig. 4.9. Daily repeatabilities of coordinates for the CRODYN'94 campaign, derived by estimating 7 Helmert transformation parameters each day using the free solutions

CRODYN'96 Day	RMS (in M) [Single Difference]	No. of Observations	No. of Parameters	No. of Coordinates
253	0.003	591161	586	90
254	0.003	606569	599	90
255	0.004	511890	538	84

Table 4.5. Characteristics of the adjustment for the daily solutions within the CRODYN'96

Campaign Solution	RMS (in M) [Single Difference]	No. of Observations	No. of Parameters	No. of Coordinates
CRODYN'94	0.003	1458167	971	73
CRODYN'96	0.003	1709620	1552	93

Table 4.6. Characteristics of the adjustment for the free campaign solutions within the CRODYN'94 and CRODYN'96 campaigns

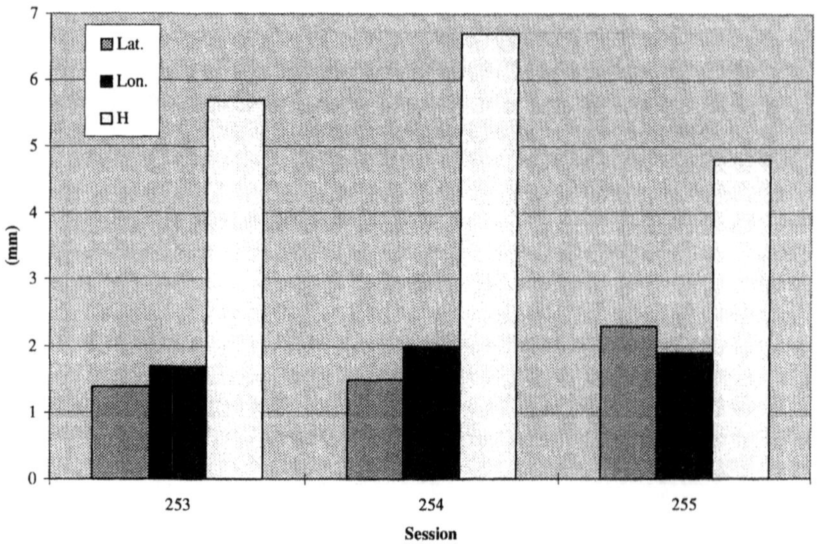

Fig. 4.10. Daily repeatabilities of coordinates for the CRODYN'96 campaign, derived by estimating 7 Helmert transformation parameters each day using the free solutions

The normal equations from each daily solution were combined using the ADDNEQ program of the Bernese software to create a free campaign solution with 1 m a priori standard deviations for all station coordinates (*Table 4.6*). On the basis of a 7 Helmert transformation parameters, the station coordinates of the daily solutions of each individual campaign were compared separately with the results of the free campaign solution. The standard deviations for the station coordinates derived by the 7 Helmert transformation parameters are less than ±3 mm for the horizontal components and approx. ±5 mm for the vertical component (*Fig. 4.9 and 4.10*).

4.5.5 Multi Campaign Solutions

The velocities and mean coordinates of the stations were determined using ADDNEQ by combining the normal equations from each campaign (Brockmann 1996). Assumed that all estimated parameters of the campaigns have the same quality and are uncorrelated, the observations of the coordinate components y can be modelled as a linear function of time $y + e = at + b = X\text{ß}$ with $D(y) = \sigma_0^2 I$ (Brockmann 1996). The unknown parameters $\text{ß} = (a, b)^T$ can be computed according to the *Gauß–Markoff Model* ($X^T P X \widehat{\beta} = X^T P y$) with its variance–covariance matrix $D(\widehat{\beta}) = \widehat{\sigma}^2 (X^T P X)^{-1}$, with n number of observations, u number of unknown parameters, β vector of unknows (ux1), X matrix of coefficientes with full rank rgX (nxu), y vector of observations (nx1), e vector of residuans (nx1), P positive definite weight matrix, σ^2 variance of unit weight. In this case, the modelling of y as a linear function of time can be interpretated as a model of linear regression.

For the computation of the accuracy for the station coordinates the transformation of the Cartesian error components to the ellipsoidal components were made by means of the position vector ($x_{i.}$)

$$x_{i.} = [(c/V + H) \cos B \cos \lambda, \ (c/V + H) \cos \phi \sin \lambda, \ (b/V + H) \sin \phi], \tag{4.5–1a}$$

and the coefficient matrix $X = c^a_{i.} = \partial q^a \,/\, \partial x_{i.}$ for $q^a = (\lambda, \phi, H)$, using the formulas

$$X = \begin{bmatrix} -\sin\lambda \,/\, N_1 & \cos\lambda \,/\, N_1 & 0 \\ -\cos\lambda \sin\phi \,/\, N_2 & -\sin\lambda \sin\phi \,/\, N_2 & \cos\phi \,/\, N_2 \\ \cos\lambda \, cos\phi & \sin\lambda \cos\phi & \sin\phi \end{bmatrix} \tag{4.5–1b}$$

with $N_1 := cos\phi(R_1 + H)$, $N_2 := (R_2 + H)$ and a, b, c, R_1, R_2, V according to (*2.3–12a*).

In the inverse situation, for the transformation of the ellipsoidal error components to the Cartesian components the coefficient matrix $X = c_{i.a} = \partial x_{i.} \,/\, \partial q^a$ were determined with

$$X = \begin{bmatrix} -\sin\lambda \, N_1 & -\cos\lambda \sin\phi \, N_2 & \cos\lambda \, cos\phi \\ \cos\lambda \, N_1 & -\sin\lambda \sin\phi \, N_2 & \sin\lambda \cos\phi \\ 0 & \cos\phi \, N_2 & \sin\phi \end{bmatrix}. \tag{4.5–1c}$$

All station coordinates computed were referred to the mean epoch 1995.6 (ITRF94). Two different multi campaign solutions were performed to describe the velocity field of the investigation area and to compute internal and external deformation measures. The standard deviations of the estimated velocities are of the order of about ±2 mm/yr for the horizontal components and about ±4 mm/yr for the vertical component.

4.5.5.1 Absolute Coordinate Solution

For the computation of the absolute station coordinates as well as the velocity field in the network 4 IGS stations were introduced in the processing. The orbits used in the processing of the data collected within the CRODYN'94 and CRODYN'96 campaigns refer to the ITRF92 for the CRODYN'94 and to the ITRF94 for CRODYN'96. Therefore, the absolute coordinates of the IGS stations were computed in the different realization of the ITRF. The determination of the velocities by the absolute model, illustrated in *Fig. 4.11 and 4.12*, were made through the fixing of the coordinates and velocities of the 4 IGS stations [Wettzell (Germany), Matera (Italy), Graz (Austria) and Zimmerwald (Switzerland)]. In this case, the results give the velocities of the stations within the CRODYN network relative to the fixed IGS stations on the Eurasian plate in the ITRF94, epoch 1995.6. The magnitudes of the horizontal velocities vary between 1.5 and 2.5 cm/yr and show a north–east direction. The value of the height changes is in the order of 1 cm/yr.

4.5.5.2 Relative Coordinate Solution

In the relative method, the station's velocities were determined relative to the fixed coordinates and velocities of the station Graz, which is the IGS station closest to the CRODYN network. The results of the velocities obtained by the relative method are illustrated in *Fig. 4.13 and 4.14*.

The magnitudes of the horizontal velocities relative to the station Graz are in the range from 0.5 and 1.5 cm/yr, increasing from north to south. The stations in the southern part indicate a north–south direction while those in the northern part turn more to the west. The changes of the vertical components vary between 0.5 and 2 cm/yr. In this case, a rising of the area have been observed.

4.5.6 Internal Surface Deformations

As the velocities obtained by the absolute model depend on the motions assumed for the fixed stations, only the results of the relative model were taken into consideration for the determination of deformation. In order to determine the internal and external deformation measures for the investigation area, an area–wide grid was defined on the basis of the existing observation points with a grid increment of 0.1° in the region between 42.4° and 46.4° in latitude and from 13.6° to 17.6° in longitude.

The interpolation of the heights, horizontal and vertical velocities, as well as the required derivatives of the heights and velocities with respect to the surface coordinates u^α for the grid points were determined using the collocation method.

For surface dilatation the sign of q indicates the nature of the deformation. A positive sign indicates the regions where distances become longer (regions of extension), a negative sign describes the regions where distances become shorter (regions of compression).

The results of deformation analysis based on the analytical surface deformation theory, described in *sections 2.3.2 and 2.3.3*, indicate three different deformation zones (*Fig. 4.15*). The maximum deformation was observed in the north–western part. This area is a dominant extension area. The values of the extensions are in the order of 2 mm/yr per 10 km. The results of the surface elongation of this area lead to extensions as shown in *Fig. 4.16.* In this part, a south–northwest and northwest–southeast extension has been found.

Another main extension area can be seen in the eastern and in the south–eastern part of the investigation area. There, the extensions are smaller than in the north–western part. In this area, the values of the extensions amount to 0.5–1.0 mm/yr per 10 km, and show a south–northwest direction.

Between the two extension areas, a compression area has been observed (*Fig. 4.15*). The values of the compression are in the order of 1–2 mm/yr per 10 km, and north–south and north–southeast compression can be seen. The north–eastern part of the investigation area can be assumed as stable.

4.5.7 External Surface Deformations

The external surface deformations are described by means of changing of the mean curvatures (δH) according to (2.3–21c) which are computed as changes of the first and second fundamental tensor (2.3–17a), and the results are represented as isolines in *Fig. 4.17.* The maximum changes of the mean curvature were observed in the north–western and in the southern part of the investigation area. The north–western part is a sinking area, and there, the mean Earth's radii are decreasing. Another sinking area can be seen in the surroundings of Split. The southern part of the investigation area, in the surroundings of Dubrovnik, is a rising area. There, the mean Earth's radii are increasing.

Prof. Dr. Krešimir Čolić (right), *Geodetic Faculty of the University of Zagreb*, and Dipl.–Ing. Zlatko Medić *Head of the Department Land Surveying of the State Geodetic Administration in Zagreb* are controlling the eccentricity of Antenna at station Makarska during the CRODYN'98 campaign

4.5.8 Figures of the Adriatic Sea Area

HORIZONTAL VELOCITIES RELATIVE TO THE IGS STATIONS

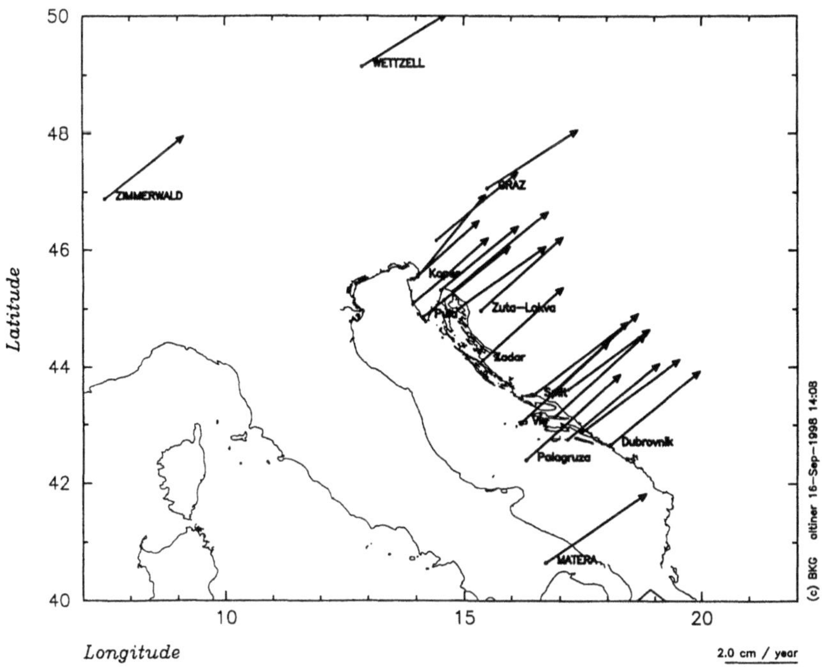

Fig. 4.11. Horizontal velocities computed within the CRODYN'94 and CRODYN'96 GPS campaigns. Coordinates of the IGS stations Wettzell, Graz, Matera and Zimmerwald were fixed

VERTICAL VELOCITIES RELATIVE TO THE IGS STATIONS

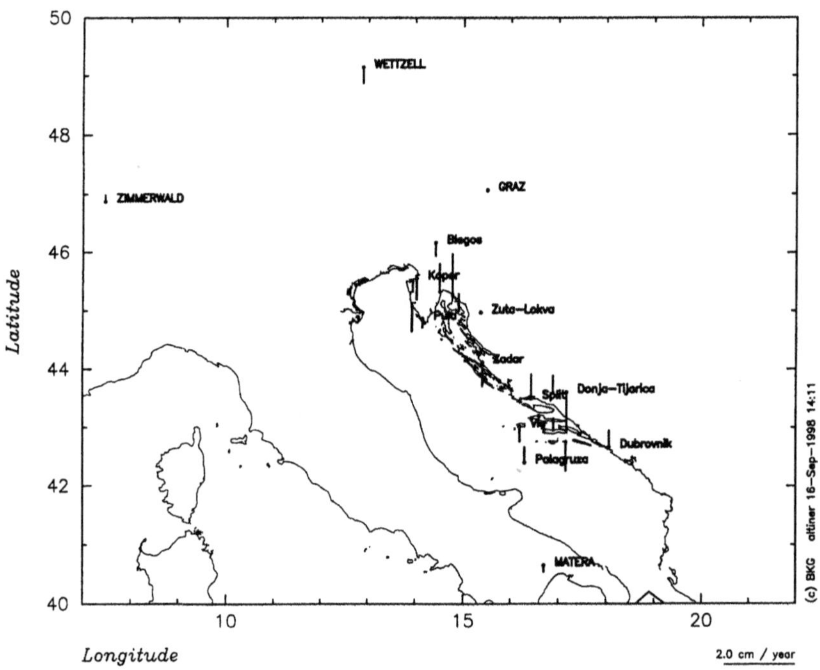

Fig. 4.12. Vertical velocities computed within the CRODYN'94 and CRODYN'96 GPS campaigns. Coordinates of the IGS stations Wettzell, Graz, Matera and Zimmerwald were fixed

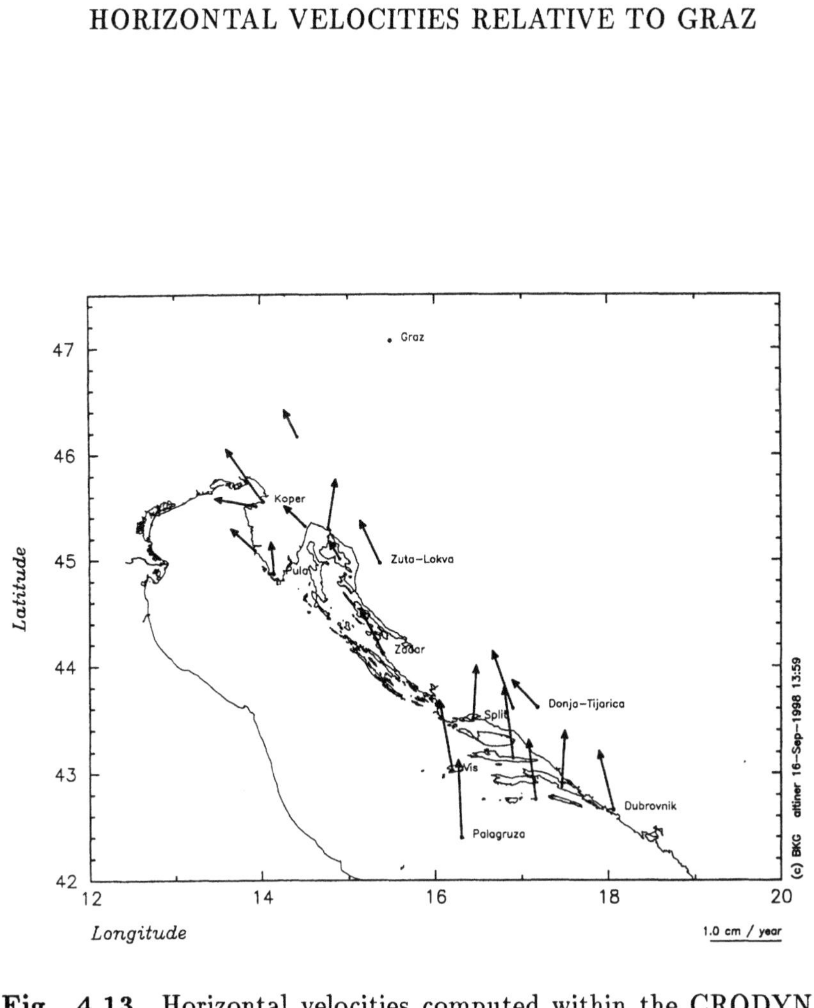

Fig. 4.13. Horizontal velocities computed within the CRODYN'94 and CRODYN'96 GPS campaigns. Coordinates of the IGS station Graz were fixed

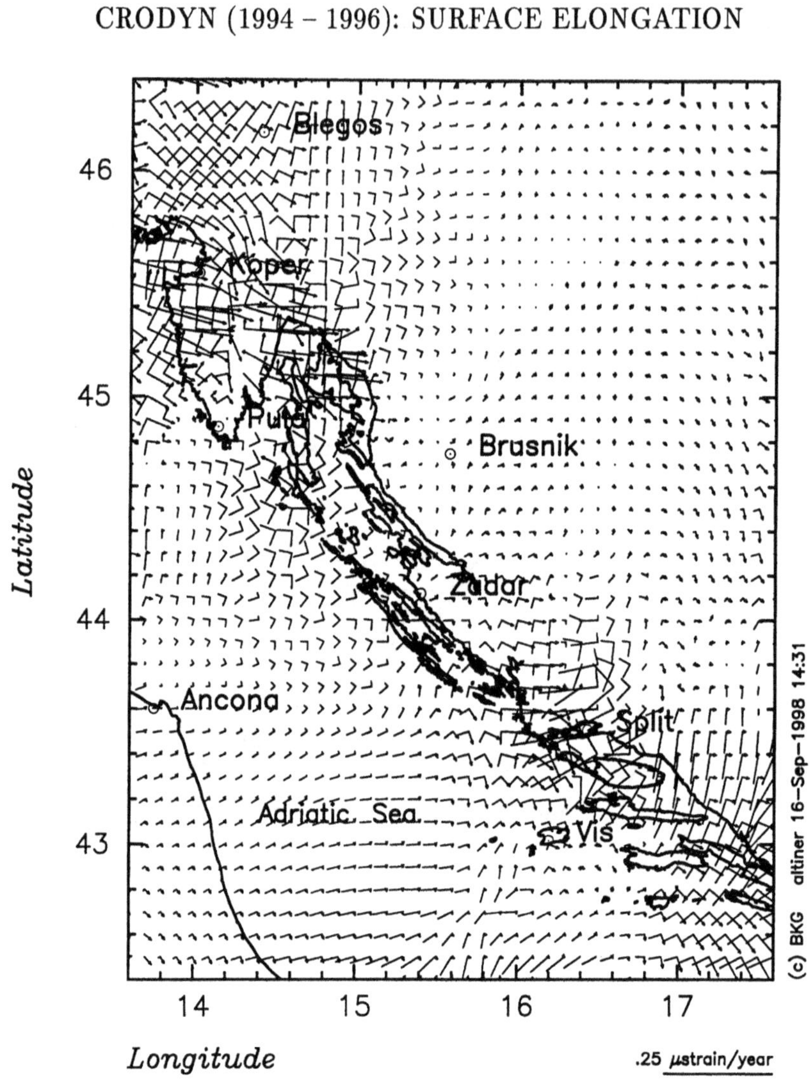

Fig. 4.16. Surface elongations. The arrows indicate the direction and magnitude of extension. The lines show the direction and magnitude of compression. The results are given in *µstrain*

CRODYN (1994 – 1996): CHANGE OF MAIN CURVATURE

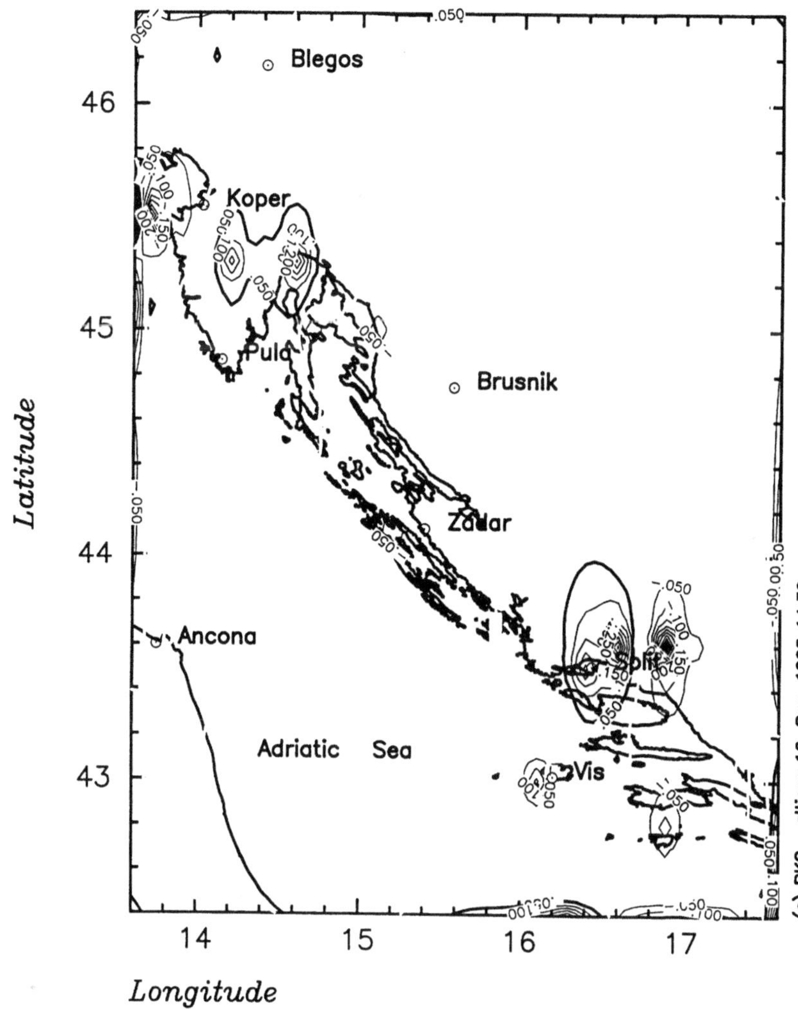

Fig. 4.17. Changes of mean curvatures. The results shown are multiplied by 10^7. A positive sign shows the region of sinking, and a negative sign means the region of rising

4.6 Summary

The frequencies of earthquakes occurences in the Adriatic Sea area show that the seismic activity is concentrated especially in the northwestern part and southern part of the investigation area. The last earthquakes in Croatia and Slovenia in 1996 and in 1998 can be seen as supporting this statement. A comparison of the results derived on the basis of the analytical surface deformation theory with the results of the seismological and geological investigations in the same area indicates a good coincidence between two results. The concentration of the earthquakes in the Adriatic Sea area covers the area with the maximum surface elongations computed within the CRODYN'94 and CRODYN'96 GPS campaigns.

The maximum deformations derived in the north–western part of the investigation area with an extension of 2 mm/yr per 10 km in the north–western and south–eastern directions could be seen as an indication of the magnitude 5.8 earthquake in Bovec (Slovenia) on April 12, 1998. A magnitude 6.0 earthquake in Ston occurred on September 5, 1996 in the southern Croatia, also an area where maximal extensions was observed.

Preliminary results obtained from two GPS campaigns support the opinion that the Adriatic Sea is a promontory of the African plate (*Fig. 4.15 and 4.16*). A south–north compression of 1 to 2 mm/yr per 10 km across the regions of extensions is consistent in direction and magnitude with the movement of the African plate in this area (Mantovani et al. 1992).

Concidering the short duration of the investigation period, the question of whether the Adriatic Sea is a promontory of the African plate or exists as an independent microplate cannot be answered seriously. Moreover, for the elimination of the model errors and influences of local geological processes on the station's coordinates and velocities, the results of a third GPS campaign are to be taken into consideration. Furthermore, for a seriously assertion about the present tectonic activities in the Adriatic Sea area by means of the GPS, a long investigation period from 8 to 10 years is to be covered.

Bibliography

Abramowsk A, Müller H (1992) Geometrisches Modellieren. Bibliographisches Institut, Mannheim Wien Zürich

Altıner Y (1992) SLR– und GPS–Messungen in Anatolien und erste Resultate. Vermessungswesen und Raumordnung, pp 393–399

Altıner Y, Seeger H (1993) Is the Motion of the Eastern Mediterranean Region Faster than Expected?. Geological Journal Vol. 28, pp 319–325

Altıner Y, Colic K, Gojceta B, Lipej B, Marjanovic M, Rasic L, Seeger H (1995) Results of the EUREF–1994 Croatia and Slovenia GPS campaign. In: Gubler E, Hornik H (eds) Report on the Symposium of the IAG Subcommission for EUROPE (EUREF). Astronomisch–Geodätische Arbeiten No. 56, München, pp 51–57

Altıner Y (1996) Geometrische Modellierung innerer und äußerer Deformationen der Erdoberfläche mit Anwendungen an der Nordanatolische Verwerfung und in der West–Türkei. Deutsche Geodätische Kommission, C 462 Mitteilungen des Instituts für Angewandte Geodäsie, Frankfurt am Main

Altıner Y, Basic T, Colic K, Gojceta B, Marjanovic M, Medic Z, Rasic L, Seeger H (1997a) Results of the CROREF'96 GPS Campaign. In: Gubler E, Hornik H (eds) Report on the Symposium of the IAG Subcommission for EUROPE (EUREF). Astronomisch–Geodätische Arbeiten No. 58, München, pp 108–123

Altıner Y, Miskovic D, Seliskar A, Seeger H, Tavcar, D (1997b) Results of the SLOVENIA'95 GPS Campaign. In: Gubler E, Hornik H (eds) Report on the Symposium of the IAG Subcommission for EUROPE (EUREF). Astronomisch-Geodätische Arbeiten No. 58, München, pp 124–132

Altıner Y (1998) Nutzung von GPS-Beobachtungen in der Ebene und ein neues Verfahren zur Darstellung von äußeren Flächenänderungen, Vermessungswesen und Raumordnung, Bonn, pp 42–50

Anderson H, Jackson J (1987) Active tectonics of the Adriatic region. Geophys. J. R. Astr. Soc. 91, pp 937–983

Argus D F, Gordon R G (1991) No-Net-Rotation model of current plate velocities incorporating plate rotation model NUVEL-1, Geophys. Res. Lett., 18, pp 2039–2042

Ayan T, Çelik R N, Alanko M G, Denli H, Özlüdemir M T, Groten E, Leinen S, Traiser J (1997) Preliminary results of deformation measurements on Karasu Viaduct using GPS technique. In: Altan O, Gründig L (eds) Second Turkish-German Joint Geodetic Days, Berlin, pp 37–50

Babbucci D, Tamburelli C, Mantovani E, Albarello D (1997) Tentative list of major deformation events in the Central-Eastern Mediterranean region since the middle Miocene. Anali Di Geofisica, Vol. XL, Nr. 3

Bilajbegovic A (1996) Erforschung der Verschiebung des Staudammes vom Wasserkraftwerks PERUCA mit den klassischen und modernen elektronischen Instrumenten. In: Pelzer H, Heer R (eds) Proceedings of the 6th International FIG-Symposium on Deformation Measurements, Wissenschaftliche Arbeiten der Fachrichtung Vermessungswesen der Universität Hannover, Nr. 217

Bock Y (1982) The use of the baseline measurements and geophysical models for the estimation of crustal deformations and the terrestrial reference system. The Ohio State University Research Foundation Columbus, Ohio 43212

Boucher C, Altamimi Z, Duhem L (1992) ITRF91 and its assoc velocity field. IERS Technical Note 12, Observatoire de Paris

Boucher C, Altamimi Z, Duhem L (1993) ITRF92 and its assoc velocity field. IERS Technical Note 15, Observatoire de Paris

Boucher C, Altamimi Z, Duhem L (1994) Results and Analysis of the ITRF93. IERS Technical Note 18, Observatoire de Paris

Brockmann E (1996) Combination of Solutions for Geodetic and Geodynamic Application of the Global Positioning System. Ph.D. Thesis, Astronomical Institute, University of Berne Switzerland

Brunner F K (1979) On the Analysis of Geodetic Networks for the Determination of the Incremental Strain Tensor. Survey Review, pp 56–67

Caspary W (1987) Concepts of Network and Deformation Analysis. School of Surveying, The University of New South Wales

Celet P (1977) The dinaric and Aegean arcs: the geology of the Adriatic. In: Nairn A E M, Kanes W H, Stahli F G (eds) The Ocean Basins and Margins 4A, pp 215–261

Channel J E T, D'Argio B, Horvath F (1979) Adria, the African Promontory, in Mesozoic Mediterranean Paleogeography, Earth's Science Rev. 15, pp 213–292

Conzett R, Matthias H J, Schmid H (1980) Beiträge zum VIII. Internationalen Kurs für Ingenieurvermessung 1980, Dümmler, Bonn

Cui J, Freeden W, Witte B (1992) Gleichmäßige Approximation mittels sphärischer Finite–Elemente und ihre Anwendung in der Geodäsie. Zeitschrift für Vermessungswesen, pp 266–278

D'Argio B, Horvath F (1984) Some remarks on the deformation history of Adria, from the Mesozoic to the Tertiary. Ann. Geophys., 2 (2), pp 143–146

DeMets C, Gordon R G, Argus D F, Stein S (1990) Current plate motions. Geophysical Journal Int. 101, pp 425–478

DeMets C, Gordon R G, Argus D F, Stein S (1994) Effects of revisions, to the geomagnetic reversal time scale on estimates of current plate motions, Geophys. Res. Lett., Vol. 21, pp 2191–2194

Drewes H (1982) A geodetic approach for the recovery of global kinematic plate parameters. Bull. Géod. 56, pp 70–79

Drewes H (1993) A Deformation Model of the Mediterranean from space Geodetic Observations and Geophysical Predictions. Springer, IAG symposia, Vol. 112, pp 373–378

Duschek A, Hochrainer A (1965–1968) Tensorrechnung in analytischer Darstellung. Springer, Wien New York

Encarnação J L, Hoschek J, Rix J (1990) Geometrische Verfahren der graphischen Datenverabeitung. Springer, Berlin, Heidelberg, New York

Falkenberg T (1993) Genauigkeitsmaße in geodätischen Netzen und ihre elastostatischen Analogien. Veröff. des Geodätischen Instituts der Technischen Universität München

Flügge W (1972) Tensor analysis and continuum mechanics. Springer, Berlin, Heidelberg, Newyork

Geiß E (1987) Die Lithosphäre im Mediterranean Raum, Ein Beitrag zu Structur, Schwerefeld und Deformation. Deutsche Geodätische Kommission, C 32, München

Gerstenecker C, Demirel H (1993) Deprem kestirimi araştırmalarında jeodezik katkılar. 4. Harita Kurultayı, Ankara, pp 203–215

Giese P, Reutter K–J (1978) Crustal and structural features of the margins of the Adria Microplate. In: Closs H, Roeder D, Schmidt K Alps, Apennines, Hellenides. Stuttgart, pp 565–588

Ghitău D (1998) Beiträge der Geodäsie zur Beschreibung der Zustandsänderung eines deformierbaren Körpers im Nachbereich. Allgemeine Vermessungsnachrichten, pp 239–246

Grafarend E (1977) Stress–strain Relations in Geodetic Networks. Communication from the Geodetic Institute Uppsala University No. 16

Hauck H, Reinhart E, Wilson P (1992) Das WEGENER-MEDLAS-Projekt. Zeitschrift für Vermessungswesen, pp 195–205

Heck B (1984) Zur geometrischen Analyse von Deformationen in Lagenetzen. Allgemeine Vermessungsnachrichten, pp 357–364

Heitz S (1968) Geoidbestimmung durch Interpolation nach kleinsten Quadraten aufgrund gemessener und interpolierter Lotabweichungen. Deutsche Geodätische Kommission, C 124, Frankfurt am Main

Heitz S (1980–1983) Mechanik fester Körper, Dümmler, Bonn

Heitz S (1988) Coordinates in Geodesy. Springer, Berlin Heidelberg New York

Heitz S, Stöcker–Meier E (1998) Grundlagen der Physikalischen Geodäsie, Dümmler, Bonn

Hekimoğlu Ş (1997) Finite Sample Breakdown Points of Outlier Detection Procedures. Journal of Surveying Engineering, vol 23(1), pp 15–31

Gurtner W, Mader G, MacArthur D (1989) A common Exchange Format for GPS Data. In: Proceedings of the Fifth International Geodetic Symposium on Satellite Systems, Las Cruces, New Mexico

Iz H B (1987) An Algorithmic Approach to Crustal Deformation Analysis. Rep. Ohio State Univ. Nr. 382

Jackson J, McKenzie D (1988) The relationship between plate motions and seismic moment tensors, and the rates of activa deformation in the Mediterranean and Middle East. Geophysical Journal (1988) 93, pp. 45–73

Kaniuth K, Kleuren D, Tremel H, Schlüter W (1998) Elevationabhängige Phasenzentrumsvariationen geodätischer GPS–Antennen. Zeitschrift für Vermessungswesen, pp 319–325

Kent C C (1976) Plate Tectonics and Crustal Evolution. Pergamon Press, Inc., New York Toronto Oxford Sydney Braunschweig Paris

Koch K R (1985) Ein statistisches Auswerteverfahren für Deformationsmessungen. Allgemeine Vermessungs–Nachrichten, pp 97–108

Koch K R (1988) Parameter Estimation and Hypothesis Testing in Linear Models. Springer, Berlin Heidelberg New York

Köhler M (1986) Ein geodätischer Beitrag zur Erfassung und Darstellung des Verzerrungsverhaltens von Eisflächen unter Anwendung der Kollokationsmethode. Deutsche Geodätische Kommission, C 318, München

König R, Weise K H (1951) Mathematische Grundlagen der Höheren Geodäsie. Springer, Berlin Göttingen Heidelberg

McKenzie D (1972) Active Tectonics of the Mediterranean Region. Geophysical Journal of the Royal Astronomical Society 30, pp 109–185

Mantovani E, Albarello D, Babbucci D, Tamburelli C (1992) Recent Geodynamic Evolution of the Central Mediterranean Region Tottonian to Present. Department of Earth Sciences–University of Siena

McCarthy D D (1996) IERS Technical Note 21, IERS Convensions (1996), Observatoire de Paris

Means W D (1976) Stress and strain, Basic Concepts of Continuum Mechanics for Geologists. Springer, Newyork Heidelberg Berlin

Meier S, Keller W (1990) Geostatistik. Akademie–Verlag, Berlin

Melchior P (1983) The Tides of the Planet Earth. Pergamon, Oxford New York Frankfurt am Main

Mervart L (1995) Ambiguity Resolution Techniques in Geodetic and Geodynamic Application of the Global Positioning System. Inauguraldissertation der Philosophisch–Naturwissenschaftlichen Fakultät der Universität Bern

Milev G (1992) Geodätische Methoden zur Untersuchung von Deformationen, Konrad Wittwer Verlag, Stuttgart

Miller H (1992) Abriß der Plattentektonik. Enke, Stuttgart

Mišković D, Altıner Y (1997) National Report of the Republic of Slovenia. In: Gubler E, Hornik H (eds) Report on the Symposium of the IAG Subcommission for EUROPE (EUREF). Astronomisch–Geodätische Arbeiten No. 58, München, pp 108–123

Moritz H (1973) Least Squares Collocation. Deutsche Geodätische Kommission, A 75, München

Moritz H (1980) Advanced Physical Geodesy. Wichmann, Karlsruhe und Abacus Press, Tunbridge Wells Kent

Moritz H, Sünkel H (1977) Approximation Methods in Geodesy. Wichmann Verlag, Karlsruhe

Müller I I, Serbini S (1989) The Interdisciplinary Role of Space Geodesy. Lecture Note in Earth Sciences, Band 22, Springer–Verlag Berlin Heidelberg New York London Paris Tokyo Hong Kong

Niemeier W (1976) Grundprinzipien und Rechenformeln einer strengen Analyse geodätischer Deformationsmessungen. In: Eichhorn G, Kobold F, Rinner K (eds) Beiträge zum VII. Internationalen Kurs für Ingenieurvermessung 1976. Schriftenreihe Wissenschaft und Technik der Technischen Hochschule Darmstadt

Niemeier W (1985) Deformationsanalyse. In: Pelzer H (ed) Geodätische Netze in Landes– und Ingenieurvermessung. Konrad Wittwer Verlag, Stuttgart

Noomen R, Springer T A, Ambrosius B A C, Herzberger K, Kuijper D C, Mets G–J, Overgauw B, Wakker K F (1996) Crustal Deformations in the Mediterranean Area Computed From SLR and GPS Observations. J. Geodynamics Vol. 21, No. 1, pp 73–96

Öztürk E, Şerbetçi M (1989) Dengeleme Hesabı II. Karadeniz Teknik Üniversitesi Mühendislik Fakültesi, Trabzon

Pelzer H (1971) Zur Analyse geodätischer Defeormationsmessungen. Deutsche Geodätische Kommission, C 164, München

Reinking J (1994) Geodätische Analyse inhomogener Deformationen mit nichtlinearen Transformationsfunktionen. Deutsche Geodätische Kommission, C 413, München

Reilinger R E, McClusky S C, Oral M B, King R W, Toksoz M N, Barka A A, Kınık I, Lenk O, Sanlı I (1997) Global Positioning System measurements of present–day crustal movements in the Arabia–Eurasia plate collision zone. Journal of Geophysical Research, Vol. 102 No. B5, pp 9983–9999

Reinhart E, Becker M (1998) Das Zentraleuropäische Geodynamikprojekt CERGOP, Mitteilungen des Bundesamtes für Kartographie und Geodäsie, Frankfurt am Main, Band 1, pp 109–120

Rothacher M, Mervart L (1996) Bernese GPS Software Version 4.0, Bern

Saastamoinen J (1972) Atmospheric Correction fot the Troposphere and Stratoshere in Radio Ranging of Satellites. Am. Geophys. Union, Geophys. Monograph Series, Vol.15, pp 247–251

Saler H (1995) Erweiterte Modellbildung zur Netzausgleichung für die Deformationsanalyse dargestellt am Beispiel der Geotraverse Venezolanische Anden. Deutsche Geodätische Kommission, C 447, München

Schneider M (1998) Earth Rotation, Reserach Group For Space Geodesy (FGS). Mitteilungen des Bundesamtes für Kartographie und Geodäsie, Frankfurt am Main, Sonderheft

Seeger H, Altıner Y, Engelhardt G, Franke P, Habrich H, Schlüter W (1998) 10 Jahre Aufbauarbeit an einem neuen geodätischen Bezugssystem für EUROPA. Mitteilungen des Bundesamtes für Kartographie und Geodäsie, Frankfurt am Main, Band 1, pp 9–51

Seno T, Morijama T, Stein S, Woods D F, DeMets C, Argus D, Gordon R (1987) Redetermination of the Philippine sea plate motion (abstract). Eos Trans. AGU, 68, p 1474

Seno T, Stein S, Gripp A E (1993) A model for the motion of the Philipinne sea plate consistent with NUVEL–1 and geological data. Journal of Geophysical Research, Vol. 98, pp 17,941–17,948

Skoko D, Mokrovic J (1998) Development of seismology – a review. In: Andrija Mohorovicic, Zagreb, pp 9–41

Sovers O J, Jacobs C S (1994) Observation Model and Parameter Partials for the JPL VLBI Parameter Estimation Software "MODEST"–1994. California, Rev. 5, JPL Publication 83–39

Sperling D (1994) Zeitabhängige Gravitationseffekte am Beispiel des Pumpspeicherkraftwerkes Vianden. Institut für Theoretische Geodäsie der Universität Bonn

Taymaz T, Eyidoğan H, Jackson J (1991) Source parameters of large earthquakes in the East Anatolian Fault Zone (Turkey). Geophys. J. Let. 106, pp 537–550

Udias A (1982) Seismicity and Seismotectonic Stress Field in the Alpine–Mediterranean Region. Am. Geophys. Union, Geodynamics Series, Vol. 7, pp 75–82

Vandenberg J, Zijderveld J D A (1982) Oalemagnetism in the Mediterranean area. In: Berkhemer H, Hsu K J (eds) Alpine Mediterranean Geodynamics. Am Geophys. Un. Geodyn. Ser., 7, pp 83–112

Vaniček P, Krakiwsky E (1986) Geodesy, the concepts. Elsevier Amsterdam Lausanne New York Oxford Shannon Tokyo

Weber G, Becker M, Franke P (1998) GPS–Permanentnetze in Deutschland und in Europa. Mitteilungen des Bundesamtes für Kartographie und Geodäsie, Frankfurt am Main, Band 1, pp 93–108

Westaway R (1993) Quaternary Uplift Southern Italy. Journal of Geophysical Research, Vol. 98, No. B2, pp 21,741–21,772

Welsch W (1989) Strainanalyse aus geodätischen Netzbeobachtungen. In: Kersting N, Welsch W Rezente Krustenbewegungen. Schriftenreihe Vermessungswesen der Universität der Bundeswehr No. 39, München

Wilmes H (1983) Entwicklung eines horizontalen Stabextensometers mit hydrostatischer Lagerung zur Messung lokaler Gezeitendeformationen der Lithosphäre. Deutsche Geodätische Kommission bei der Bayerischen Akademie der Wissenschaften, C 281, München

Wilson D S (1993a) Confirmation of the astronomical calibration of the magnetic polarity timescale from see–floor spreading rates. Nature, 364, pp 788–790

Wilson D S (1993b) Confidence intervals for motion and deformation of the Juan de Fuca Plate. Journal of Geophysical Research, Vol. 98, pp 16,053–16,071

NIMA (1997) World Geodetic System 1984, Its Definition and Relationships with Local Geodetic Systems. Technical Report of the National Imagery and Mapping Agency of the Department of Defense (USA)

Wübbena G, Schmitz M, Menge F, Seeber G, Völksen C (1997) A new approach for Field Calibration of Absolute Antenna Phase Center Variations. Navigation 44 (2), pp 247–255

Zienkiewicz O C (1972) The Finit Element Method. MacGraw Hill Book Campany (UK) Linited Maidenhead–Berkshire–England

Index

9 783540 65820

GPSR Compliance

The European Union's (EU) General Product Safety Regulation (GPSR) is a set of rules that requires consumer products to be safe and our obligations to ensure this.

If you have any concerns about our products, you can contact us on ProductSafety@springernature.com

In case Publisher is established outside the EU, the EU authorized representative is:

Springer Nature Customer Service Center GmbH
Europaplatz 3
69115 Heidelberg, Germany

Batch number: 10584448

Printed by Printforce, the Netherlands